THE
MONSANTO
PAPERS

THE
MONSANTO
PAPERS

Corruption of Science and
Grievous Harm to Public Health

GILLES-ÉRIC SERALINI
WITH JÉRÔME DOUZELET

Foreword by Vandana Shiva
Translated from the French by Lucinda Karter

Skyhorse Publishing

Children's
Health Defense

Skyhorse Publishing books may be purchased in bulk at special discounts for
sales promotion, corporate gifts, fund-raising, or educational purposes. Special
editions can also be created to specifications. For details, contact the Special
Sales Department, Skyhorse Publishing, 307 West 36th Street, 11th Floor,
New York, NY 10018 or info@skyhorsepublishing.com.

Skyhorse® and Skyhorse Publishing® are registered trademarks of Skyhorse
Publishing, Inc.®, a Delaware corporation.

Visit our website at www.skyhorsepublishing.com.

10 9 8 7 6 5 4 3 2 1

Library of Congress Cataloging-in-Publication Data is available on file.

Cover design by Brian Peterson

Print ISBN: 978-1-5107-6763-8
Ebook ISBN: 978-1-5107-6764-5

Printed in the United States of America

Contents

Contents

Foreword

**Defending Our Food Supply, Our Health, Our Rights,
but Also the Integrity of Science**

Rachel Carson's *Silent Spring* (1962) awakened us to the dangers of pesticides. In the decades following its publication, regulations were put in place and scientific research evolved on the dangers to our health of pesticides, and, more recently, the biosafety of GMOs. I used to be a member of the United Nations committee of experts on biosecurity following article 19.3 of the Convention on Biosecurity. This is around the time when the attacks against science and scientists began.

In the decades that followed, the word "science"—which comes from the Latin root scire, "to know"—has seen its meaning eviscerated, stripped down to its bare bones, and reduced to nothingness. The pursuit of knowledge has been replaced by the discourse of the paragons of the "defenders of science," their mouths lined with dollar bills. Their baseless knowledge, supposedly extracted from "facts," has been wrongfully accepted as a panacea for the world's very real population issues.

This book by Dr. Gilles-Éric Seralini, a renowned scientist, researcher, and toxicologist, and Jérôme Douzelet, a celebrated chef, gardener, and expert in "Real Food," is a wake-up call to the dangers of the toxicities in our food. It is also a dire warning about the growing threat to public health and security, science and knowledge, freedom and democracy. It's what happens when propaganda replaces fact and when the Cartel of Toxics, led by

Monsanto, now Bayer, begins to take control of our food supply, our public policy, our regulatory agencies, and the media.

Allowing toxics to dominate our food and our agricultural industries is effectively now supported by putting a stronghold on the very foundations of science and education.

We know these chemicals are toxic. We have come to recognize that the companies peddling pesticides deny their dangers in order to pursue their quest for profit to the detriment of life and health.

We need real science and scientists who have no ties to or interests in the food industry when we evaluate the effect of chemicals and GMOs on our food supply. It is nothing less than the very protection of our public health.

It is for these reasons that today more than ever, it has become a matter of life and death that we guarantee the independence of science, as well as the freedom and integrity of scientists.

It is an honor to pen the preface to this book. Dr. Gilles-Éric Seralini is a brilliant researcher who succeeded in upholding the integrity of science in the face of Monsanto's attacks. This book is a unique testimony for our times. In the end, it was proven that the accusations brought against him by Monsanto were fabricated, and that there was overwhelming proof of a link between Roundup and various types of cancer.

This important book chronicles the Toxic Cartel's crimes and Dr. Seralini's dedication to preserving the honesty and independence of his research. It also awakens us to the very real necessity of protecting our knowledge, our health, and our freedom into the future against the giants of agrobusiness and their chemical toxins, and the titans of the pharmaceutical and technology industries. Indeed, taken together, they aim for a stronghold on our agriculture, our food, our public health, and our scientific community.

To me, science is a quest for truth, a search for models, and relations between things, especially in the realm of living things. It is an epistemic process, an open conversation, a dialog between nature and society in order to further knowledge through interconnectivity, integrity, discussion, debate, openness of mind, transparency, and accountability.

These are the reasons why I decided to become a scientist and physicist, and later, just as disasters were unfolding in Punjab and Bhopal, India, I went on to study the ecology of agriculture and food.

This is when I became aware of the fact that agriculture and all the life sciences had been co-opted by the Poison Cartel, which, after all, had first begun manufacturing deadly chemicals for use in Hitler's concentration camps. Lumped together under the name IG Farben, these companies worked side by side with US corporations during the war. Even as their armies fought one another, transatlantic industrial alliances were formed, such as the one between Standard Oil and IG Farben,[1] or MoBay (Monsanto and Bayer) to research new ways to manufacture poisons initially designed to kill human beings.

The scientists and management at the Chemistry Cartel IG Farben were tried in Nuremberg for crimes against humanity. But despite the Nuremberg trial, the Poison Cartel persevered in its quest to manufacture deadly products. Once the war was over, chemical weapons became ingredients in agrochemical products. Industrial agriculture brought these into my own beloved country of India and beyond in the name of the Green Revolution.

My book *The Violence of the Green Revolution* is the result of the research I did further to the events of 1984 in Punjab.[2] Monsanto, a chemical manufacturer well into the 1980s, had just cornered the market for GMO crops and the pesticides that go with them, such as Roundup. People wrongly associate Roundup with glyphosate alone, when in fact it contains other unlisted toxins as well. The GMOs have been bred to tolerate the pesticides, so they absorb them without dying. I launched the Navdanya movement to protect copyright-free crops and to further an agriculture free of these poisons.

Professor Seralini's research highlights the differences between glyphosate and Roundup, exposing the toxicity of this latter. Needless to say, it is the bestselling toxic product in the world, used on farms, in fields, on sports terrains, walkways, public parks, and on school grounds. Once the GMO that tolerated this herbicide (pesticide is a broad term that includes herbicides, as well as insecticides and fungicides) was introduced, its use skyrocketed, increasing around fifteen-fold, from 51 million kilos in 1995 to around 750 million kilos in 2014. In all, 8.6 billion kilos of glyphosate have been spread around since it was introduced in 1974.[3]

Dr. Seralini's research shows that contrary to what Monsanto claims in its ads, and to what comes out of the mouths of its talking heads, Roundup is not safe. It not only attacks all our vital organs, but also contributes to the growth of breast tumors. Dr. Seralini and his research have been attacked, just as mine has in India on the topic of GMO cotton, also known as Bt. And the elevated cost of genetically modified crops has driven farmers into debt and to suicide. Monsanto and its scientists have denied this connection even in scientific journals and through the media.

These coordinated attacks were against actual science and independent scientists such as Dr. Seralini. His research papers proving that Roundup can cause cancer, among other illnesses, have been ascertained by the International Agency for Research on Cancer (IARC) and the World Health Organization (WHO). Worldwide, communities and even entire countries have banned Roundup and glyphosate.[4]

Thousands of cancer victims have taken justice into their own hands. Monsanto's initial attempts to hide its internal documents failed, thanks to a legal process called "discovery." This process led to the release of the "Monsanto Papers," which show clearly that the attacks against Dr. Seralini were conceived and planned out by

Monsanto. This book explains the strategy behind the Monsanto Papers to the public.

Henry Miller, quoted in the pages that follow, often criticized Dr. Seralini as well as myself. As evidenced in the Monsanto Papers, Henry Miller's articles in *Forbes* magazine were ghostwritten by Monsanto. *Forbes* had to retract these, as well as the one Miller cowrote with the shill Kavin Senapathy. Senapathy admitted that she had been used by Monsanto to promulgate this fallacy: "If you're pro-science, you must be pro-GMOs. If you're against Monsanto, then you're against GMOs. So if you're against Monsanto, you're against science." Nevertheless, as she later wrote, "I gradually realized that something about the GMO gospel sounded very wrong. In August of 2017, several articles I had co-written for *Forbes* were withdrawn when it was discovered that my co-author had lent his name to Monsanto."[5]

Over a span of three decades, I watched as Monsanto grew and tried to wield its power over the agricultural industry either through the way it licensed crops, or by forcing farmers to use GMOs, spreading its poisons. In response to this, in 2016, we put together the "The Monsanto Tribunal" along with a people's Assembly to try the corporation for crimes against nature, or "ecocides," against farmers, the people, science and scientists, and against democracy.[6]

Monsanto disappeared when Bayer acquired them in 2018. Their crimes were just beginning to be more widely recognized. Their name has changed but their misdeeds have not.[7]

In the interest of protecting life on Earth, the health of its citizens, the freedom and integrity of science and of knowledge itself, we must work together to create a future free from the attacks of "fake" science. We are currently in the grips of a food supply so artificial that it spreads ever more pandemics and illnesses.

My hope is that this important book will help us attain a democracy rooted in knowledge. That it will promote the independence of science and scientists in the essential fields of nutrition and health, as a matter of life and death. That through that very same knowledge and independence it will put an end to the dominance and infractions of the Cartel of Poison. I hope that questions of health will be based on actual facts, placed in the hands of responsible leaders of agriculture—with an agriculture that nourishes the earth and a robust food supply that feeds not only our bodies but also our spirits.

This is a call to arms, a resounding STOP to put an end to the poisoning of our bodies, to the misappropriation of science, knowledge, the media, and democracy itself.

This book invites all citizens and scientists—whoever you may be, wherever you may come from, and whatever you may do—to stand atop the heights of wisdom in order to open a new way forward toward our collective futures. The Poison Cartel—a cartel of patents—will no longer reinvent the wheel of extinction by becoming an even larger cartel of agribusiness, chemistry, biotechnology, big pharmacy, technology, and international finance. Like Monsanto, it must disappear. It's them or us. We'll be forced into extinction just like the bees and butterflies, the insects and the birds, the plants and animals being killed by products that are, essentially, manufactured for that purpose.

Every time we return to our roots, guided by the soil beneath our feet, we will get closer to harmony with ourselves and with nature. This will in turn bring us health—free from poisons, from fear of disease, pandemics, and extinction. A Return to Earth will also grow and deepen our freedom of spirit, safe from a toxic propaganda that passes for science.

The future is calling to us, and it is a future free from toxic products, thanks to this book.

—Vandana Shiva

The Seralini Affair and Monsanto's Manipulation of Science

Great is the power of steady misrepresentation; but the history of science shows that fortunately this power does not long endure.

—Charles Darwin

"The Monsanto Papers" are a set of previously confidential Monsanto documents that were released into the public domain during the course of pending toxic tort litigation, *In Re Roundup Products Liability Litigation*, 3:16-md-2741, Northern District of California. The Los Angeles law firm, Baum Hedlund Aristei & Goldman, sought the release of the documents on the grounds that none contain trade secrets that are the basis for maintaining confidentiality, and that any such claim to trade secrets must be balanced against the public health concerns surrounding Monsanto's herbicide, Roundup. The documents posted on the firm's website[1] include: internal Monsanto emails, manuscript drafts, peer review reports, deposition testimony, PowerPoint presentations, text messages, and the like.

As a research consultant to Baum Hedlund, my role was to review the confidential documents and select for de-classification the ones with the most incriminating information about how Monsanto operated. Some of these documents became exhibits used in the now-famous California trials—Johnson v. Monsanto, Hardeman v. Monsanto, and Pilliod *et al.* v. Monsanto—which led to a global US $10.5 billion settlement for plaintiffs with non-Hodgkin lymphoma induced by Roundup. These documents are now the subject of the present book.

Of particular interest is a subset of the Monsanto Papers that concern the scientific work of French toxicologist, Gilles-Éric Seralini. An article by Seralini et al., "Long Term Toxicity of a Roundup Herbicide and a Roundup-Tolerant Genetically Modified Maize," reported on a two-year feeding-study of rats in which an increase in tumors was found among rats fed genetically modified corn and the herbicide Roundup. This article was published in *Food and Chemical Toxicology* in 2012, but *after the publication of critical responses to the Seralini et al. article in 2013, the editors determined that the study was inconclusive and retracted the article.* Many of these responses were letters that called upon the editors of the journal to retract the paper. But what the Monsanto Papers reveal is the role that Monsanto played behind-the-scenes in orchestrating the retraction. One Monsanto employee reported that he had leveraged his relationship with the Editor-in-Chief of Food and Chemical Toxicology throughout the 2012 Seralini rat cancer publication and media campaign, and boasted that he had "successfully facilitated numerous third-party expert letters to the editor which were subsequently published."[2] (Seralini republished the paper in 2014 in *Environmental Sciences Europe*, 26:14, together with the explanations of the conflicts of interests *Env Sci Eur* 2014, 26:13.)

Seralini was persistent and his work presented a real threat to Monsanto's attempt to distort the science and create decoy research.

What is particularly alarming about what these documents reveal is the extent to which Monsanto was willing to go to destroy a scientist's work and career. Monsanto's campaign against Seralini also included engaging academics to attack him in the lay media and in websites that appeared to offer independent evaluations. But as it turns out, the articles were ghostwritten by Monsanto's agents and the websites were fronts secretly funded by Monsanto to defend their products.

Other documents show how Monsanto ghostwrote articles for the toxicology journals in the names of influential academics, installed industry-friendly journal editors, and manipulated peer review by having their own employees write critical reviews of articles submitted to the journals on the safety of Monsanto's Roundup, thus injecting industry bias into the process and assuring that negative assessments of Monsanto's products would be rejected.

Corporate misrepresentation of scientific testing facilitated by academics acting as shills for industry is a serious problem. When the industry uses all its resources to make sure that the scientific literature contains only their key marketing messages and suppresses any research by scientists that runs contrary to their goals, we have lost one of the most essential components of the scientific process—that it is self-correcting. Monsanto got caught meddling in the science and they paid dearly for it. Ironically and hypocritically, this was the company that used the Darwin quotation above as public relations propaganda to assure the public that their toxic chemicals were safe.

—Leemon B. McHenry
California State University, Northridge

Monsanto vs. Seralini:
Pulling Back the Curtain

In 2013, a couple of years before the World Health Organization's International Agency for Research on Cancer (IARC) analyzed Monsanto's weed killer, glyphosate, and found it to be a probable human carcinogen,[1] a technical chemical journal retracted an article[2] about a group of rats fed Monsanto's genetically modified corn and some of the company's flagship herbicide, Roundup. Until then, no animal studies longer than ninety days had been conducted using glyphosate and all the other chemicals in the herbicide's formulation that made Roundup so effective[3] at killing weeds.

Short-term trials or trials using just glyphosate alone were used to get Roundup approved and marketed. One hundred days was about the time rats or mice started to show tumors—so keeping trials at ninety days or less prevented the observation of tumor formation and enabled Monsanto[4] to market Roundup as not carcinogenic. Contrary to the accepted regulatory orthodoxy at the time, the retracted study showed otherwise—Roundup might be carcinogenic.[5]

Understandably, with billions of dollars at stake and the agribusiness industry dependent on Roundup, the orthodox scientific community appeared to rally to Roundup's defense, asserting the study was flawed, it wasn't properly peer reviewed, it used cancer-prone rats and had all sorts of other nefarious methodological and scientific defects. "Science" prevailed with the "flawed" article and its author tarred and feathered out of science town. At least that's what it looked like.

It was actually the latest of a series of concealed hit jobs against that author, Professor Gilles E. Seralini, an actual bench scientist who had the curiosity and temerity to take Monsanto's own 2004 ninety-day study used to get Roundup's regulatory approval and run it long term, for two years, which is the ordinary life-span of rats. The rats fed Roundup-laced food and water developed more tumors and other serious health deficits, suggesting Roundup wasn't as safe as it was cracked up to be. In fact, the study showed:

- 80 percent of the rats that consumed Roundup for two years developed tumors, whereas only 30 percent of the control group had tumors.
- Rats treated with Roundup had larger (30 percent to 130 percent) tumors compared to the controls.[6]

Monsanto orchestrated hit jobs in response to Seralini's findings in order to preserve Roundup's market share, or in Monsanto's terms, to maintain its "Freedom to Operate"[7] Monsanto decided to maintain its "Freedom to Operate," notwithstanding awareness among the company's own scientists that they couldn't truthfully say[8] Monsanto had established Roundup was not carcinogenic—Monsanto had purposely avoided and cleverly navigated the regulatory testing that would show the Roundup formulation's carcinogenicity.

So rather than actually conduct the long-term tests and alert, at that point, almost the entire population that the magic weed-killer and protector of genetically modified crops might be carcinogenic, Monsanto set out to suppress critics. Monsanto had different names for these suppression programs, such as "Freedom to Operate," or "FTO," "Let Nothing Go," "Whack-a-Mole," "Project Spruce," etc. The programs were conducted surreptitiously,[9] sometimes using actual corporate and retired intelligence community spies,[10] so people like Seralini and the scientific community would not know why an article got rejected, or why a published article got retracted . . . or why careers got ruined, funding dried up, jobs got lost or consumers died of non-Hodgkin's lymphoma.

The Seralini saga would likely have remained one of those unfortunate mysteries if not for the fact that a loophole in the protective orders governing internal corporate documents' confidentiality was used to unseal thousands of pages of Monsanto's internal plotting, orchestrating, and subterfuge—which became known as "The Monsanto Papers."[11]

Seralini's *Monsanto Papers* provides his insider view of what it was like to be subjected to Monsanto-instigated attacks on his studies and credibility. He walks readers through the documents themselves to show that it was Monsanto that manipulated peer reviews, engaged in ghostwriting articles that whitewashed Roundup's genotoxicity, suppressed an independent scientist's genotoxicity analysis, fed pre-written stories for reporters to "independently" publish—and even had the editor of the journal under a financial contract with Monsanto at the time the two-year study was retracted.

The documents show how Monsanto coordinated the letters-to-the-editor campaign, with scripted talking points that looked like an indignant scientific community crying foul, when it was Monsanto engaging in scientific malfeasance, violating COPE guidelines for peer review, plagiarism/ghostwriting, and conflicts of interests. One string of internal Monsanto emails in particular exemplifies why it is so important for this book and others like it to highlight the corruption of science, by financially tunnel-visioned chemical and pharmaceutical companies, that undermines the crucial safety nets we need for preventing dangerous chemicals, drugs, and vaccines getting through to unwitting patients and consumers.

In mid-2009, Seralini completed a study of liver cells exposed to Roundup, documenting liver cell damage and then showing that a treatment, DIG1, protected against cell death[12] provoked by glyphosate-based herbicides in human liver cell lines. The study showed Roundup's harm and a remedy to prevent it.

Ironically, and in gross violation of every form of conflict of interest and the fundamentals of peer-reviewed science, a Monsanto toxicologist, Bill Heydens, was asked to conduct a peer review of Seralini's paper on Monsanto's Roundup. As if that were not enough of a peer-review guidelines violation, instead of maintaining the confidentiality of the draft manuscript, Heydens circulated it to members of Monsanto's FTO team, Donna Farmer, David Saltmiras, and Steven Levine, and then synthesized their comments into what he complimented as their "excellent rebuttal material."[13]

Instead of a neutral, confidential analysis of a manuscript identifying a harm induced by Roundup and a treatment that prevented it, Heydens assembled the science-suppression team[14] to rebut the paper. They recommended rejecting publication with a number of bullet points rebutting the paper's findings and methodology. And they succeeded—the Regulatory and Toxicology and Pharmacology editor-in-chief, Gio Batta Gori, adopted "Heyden's" recommendation to reject publication. Gori passed the unfortunate news along to Seralini, without telling Seralini that one of the principal peer reviewers was Monsanto's Bill Heydens. Heydens circulated the "good news" to his FTO team and they congratulated each other on their fine accomplishment.

And thus, helpful science conveying a harm and a remedy was suppressed, and Monsanto got to maintain Roundup's Freedom to Operate, unhampered by a pesky safety publication. A few years later, after successfully engineering the tumor study retraction, the team received an award[15] for its excellent FTO work: "Achievement Title: I Smell a Rat—Response to Seralini."

Here's a description of the award:

The Seralini study was a multimedia event that was designed for maximum negative publicity. The Monsanto Toxicology Team was mobilized to provide rapid assessment of the technical aspects while the Scientific Affairs team helped organize 3rd party scientists that were fully engaged to respond to the paper. In all, there was six months of effort to respond that included Monsanto's technical evaluation, a Letter to the Editor (longer than the original manuscript), responses by the Glyphosate Task Force, powerpoint presentations, responses to numerous Regulator inquiries, blog posts and popular press articles. This was the result of coordinated efforts and synergies by people from multiple Regulatory Teams.

With no remorse or misgivings—in fact, proud of their accomplishments—the Monsanto team's efforts to combat critical thinking regarding Roundup's safety would have gone unchecked if not for publication of the "The Monsanto Papers" and the non-Hodgkin lymphoma trials transcripts—and if not for courageous scientific inquiry by individuals like Seralini and the dozens of other scientists who have deigned to present independent studies showing Roundup risks.

Indeed, $2.4 billion in jury awards in three consecutive trials, repeated losses on appeal, $10.5 billion in settlements, and the worldwide condemnation of Monsanto's conduct demonstrated by "The Monsanto Papers," and the jury trial transcripts have not resulted in a single apology, or a warning—or really any change in Monsanto's intimidation and invalidation of independent scientists such as Seralini, the IARC analysts, or the jury trial experts. Therefore, it is important to maintain the Freedom to Criticize industry-financed pseudo-science—and keep pulling the curtain back while Monsanto and its allies accuse others of the very scientific corruption they have been committing for decades.

Seralini's *The Monsanto Papers* is another strong tug in the quest to pull back that curtain.

—Michael L. Baum
Baum Hedlund
Los Angeles, California

Chapter 1
RAISING THE CURTAIN

_In which we first perceived the magnitude of the storm our
research had unleashed, and the crimes one industry had
committed._

This book tells the story of a scientist who found himself in the
midst of a murky and painful international criminal battle no
scientist should have to endure. It is written in light of the two-and-
a-half-million-page database US plaintiff lawyers obtained in 2017.
Monsanto, now Bayer, one of the largest multinational producers
of genetically modified organisms (GMOs), pesticides, and
pharmaceuticals, was compelled through dogged use of language in
the US Federal Court's order regarding the unsealing of documents
to release what became known as the Monsanto Papers.

My name appears in the database 55,952 times on 20,000 pages
covering a span of fifteen years, as hundreds of people tried to discredit my research. It was unbelievable. As an academic whose
research focused on this topic, I had become an expert-witness
throughout the world. Beginning in 2012, thousands of reporters,
public relations and marketing firms, lobbyists, pro and anti-GMO
activists, victims, and politicians from around the world had seized
upon what they called "The Seralini Affair." I would have pre-
ferred if they had called it "The Monsanto Affair" or the "The
Roundup Affair." To the public, its name evokes rats with impressive tumors, photos of which have made their way around the

world. But in reality, the company had been trying to destroy my reputation and my career ever since 2005[1] over my numerous revelations as to the hidden toxicity of its products. In the intervening years, the giant corporation was sold and lost billions in highly publicized court cases. Meanwhile, Monsanto was pushing the world's number one pesticide, Roundup, and the agricultural GMOs modified to absorb it without dying, all in the interest of simplifying industrial farming of a single crop over a very large surface, or monoculture cropping.

It was a nightmare, as much to live through as it was to painstakingly investigate the story told in this book. Our collective health and that of our planet is at stake. But I couldn't have gotten through it alone. This book was written with my friend and coauthor on three previous volumes, Jérôme Douzelet. He followed my research before it became public fodder, was up on the lawsuits that I filed, and the meetings with the American lawyers. He was with me through thick and thin, and even checked up on my health and that of my family. We attended many conferences together around the world in which we tackled these issues in front of thousands of people; the "Affair" followed us everywhere. For the purposes of this book, to simplify, we use "I" here rather than "we."

Reporters or members of the audience at talks often ask me what's become of my life as a whistleblower and what kind of threats I received or still receive as a researcher. Why do they ask, you may wonder? There are documentaries that attempt to explain it. My first reaction is to say that I've never considered myself a whistleblower but simply a researcher in the employ of the government who was doing his job, publishing his results, and, like any university professor, finding himself in need of explaining them when they reveal something new or dire. Meanwhile the threats came in all shapes and sizes, and I felt like I was living in a dictatorship.

Among my professional colleagues, I initially encountered indifference, denial, sarcasm, or my favorite, "You asked for it. You could have just reported them." This was all small potatoes compared to the discoveries I was making: Someone was committing a crime using chemicals that were deliberately minimally and poorly tested. Through it all I was like a young horse in a storm. I kept my head down and pushed on, helped along by the unwavering support of my team even as people I met in official governmental committees whispered in my ear "your career is over." After my talks, total strangers would threaten my family or threaten me physically. My superiors and colleagues begged me to remain calm lest I fly off the handle; members of my team were under all sorts of pressure at work.

From one day to the next a former president of the Parliament threatened to place me under investigation for fraud, an offense that could have cost me my job. And then the prime minister got involved (he's since been exiled in Spain) and drew my infamous rats to the attention of the president of my university, the University of Caen in Normandy. Next thing I knew, industry leaders along with their shills spoke out vehemently against me in the press, attacking my person in *ad hominem* arguments and ignoring the results of my scientific work. I started getting pummeled with insults from the powers that be, inspired by the leadership of the Institut national de la recherche agronomique (National Research Institute for Agriculture). They accused me of groundless activism, and said my papers had all been written by other people, and therefore, should be discarded. The threats increased, and I heard things like "something bad is going to happen to you." My Wikipedia profile was usurped by trolls and grossly falsified, quoting bloggers rather than my scientific references—not a surprise given the nature of the online encyclopedia. Even the crudest attack on my work could live on, fossilized like a footprint in the mudbanks of the scientific world: It made the undecided turn their backs, and it sowed doubt, its goal strategically blocking access to crucial funding for my work—work that could bring much-needed expertise to confront the marketing of toxic products that rake in billions. I was denigrated on social media along with my supporters, and I was even asked to resign from the University of Caen. Entire PR firms were hired for that purpose.

The French Academy of Science, without even bothering to ask the academics who worked for them, played the drumroll as one after the other they cast their cookie-cutter judgements on my case without asking me for my results or consulting me. While they were at it, they shared their opinions with the state of California, which was calling for an unprecedented vote on the subject of GMOs. I kept expecting the police to show up and handcuff me outside my lab. Already twenty-five police trucks were in pursuit of the demonstrators who took to the streets to support me during the first lawsuit I filed for defamation. My study and the photographs of the tumors in rats caused by Roundup, and the GMOs specifically modified to absorb it, had circled the world. In Russia and elsewhere, the GMO in question was banned, while in other parts, from Europe to Australia, my research was called into question by the press, by regulatory agencies, and the powers that be. What deity had I offended to provoke such a cataclysm? It's not like I said that the earth wasn't round, nor questioned whether man had set foot on the moon. All I had done was explain in detail the effects on our health of the most commonly used pesticide in the world, at a dose considered safe.

Exhausted and wounded to the core, like an intensive care patient on life-support, I stood my ground (I am more stubborn than most mules according to my coauthor), checking and re-checking my results, trying to block out the voices of those around me who warned me to just give up. Could it have been childish innocence to confront multibillion-dollar companies more than a little concerned by my discoveries, or the simple wish to do a good job and finish what I had started, a value inculcated in me by my parents. One thing for certain is that it stems from my desire to understand: Why cancer is so rife, along with birth defects, and lethal organic deficiencies, especially of the liver or kidneys? How can these be avoided? I was choosing to be part of something bigger than myself.

In the end, if we wanted to be as cynical as they are, we could thank Monsanto, now Bayer, for continuing their fight. Thank you for inadvertently revealing your shady operating principles not only through the Monsanto Papers, but also in the committees I served on, or even via the lobbies I fought all the way to the courts. Monsanto gives us a perfect example of how certain industry leaders, who we will uncover in the pages ahead, tried to undermine science, medicine, and public trust while they swept all critical thinking and ethics under the rug. All of this to make short term profits even as they endangered ecosystems, the climate, and world health.

We are now aware of the pure deceit with which legislation designed to protect millions of people was redirected in favor of profits and extortions, and how bloated with public funds the petrochemical industry had become. All this with the goal of getting subsidies for intensive farming, putting in place fake standards, hiding most of the toxic chemicals these products contain, and then genetically modifying crops, with the same goal in mind. Let us hope that now, more than ever, we can acknowledge a deadly precedent not to be repeated by younger generations—those young people marching in the streets to protect our climate and the planet, and to fight pesticides.

I found myself drifting away from my university. I still taught, but after this whole thing was over, I chose to continue my research with colleagues from around the world. There were several reasons for this decision. First of all, despite having excellent friends and colleagues, I could not forget the cowardly indifference I came across in meetings about the poisons being unleashed in the natural world, against our bodies and the Earth. This topic should have been of great interest in the field of education and scientific research. And then a certain few who had only published once in their lives (on the functioning of dogfish testes) started complaining about the fair division of cost of the Xerox machine. I also chose to expand the scope of my research in order to meet the most competent people around the world, and use the latest technology where it already exists, without incurring excessive costs. It's wonderful how it leads to groundbreaking collaborations on the most pressing topics, free of national or industrial restraints and with the freedom to work independently.

It's been quite a while now since universities have stopped funding projects of any originality, including those destined to shed light on the toxicity of products already authorized by nations. Luckily, nowadays foundations and group funding have taken their place, thanks to individuals who receive tax benefits for directing funds toward independent research. For the first time, thanks to globalization, we now have this original and fast way to improve global knowledge.

Rather than focus on vanishing biodiversity, university labs are (sometimes, not always) rife with the passivity of their leaders. Researchers are also subject to the fear of upsetting the system and of uninformed jealousies. For example, my thesis students were constantly finding themselves being tripped up as if banana peels had been slipped under their feet; they were paying for things I'd said on television that might have been awkward for the lab, always hopeful for some crumbs from industry that never materialized. My dear colleagues even went so far as to ban the press in our common spaces! They tried to dip into my private donations, earmarked to study the effects of pollution, to what end? To make up for the massive waste of the badly maintained pieces of equipment that didn't serve any purpose anymore, lacking trained technicians to operate them or a budget to keep them running properly. I've seen taxpayer money drowning in waste while machinery just sat there, rust forming in its circuits. But my colleagues didn't want to work with me or share materials or ideas on such a "divisive" topic. Out of pure jealousy, they repeated the things Monsanto said about me without questioning them.

The government was becoming increasingly silent on requests for research grants. Naturally this made it more difficult to work on exposing the toxicity of commercial products they had already approved. Without the liberty to freely post research topics, I had to relocate my doctoral students along with the technical aid they needed. At the same time my group was being invited to participate in international conferences around the world.

Meanwhile, I was publishing at least ten times more than what was deemed necessary to be considered an "active" researcher. More than ever, I needed to be able to report freely to my donors without passing through my slow-moving hierarchy. I needed to be free to fight my court battles without having to report back to people who had no interest in them, while at the same time, not subjecting them to the pressure of the lobbies. Concurrently, in order to get ahead and publish in major reviews, I had to keep up on the latest research methods being used abroad.

My increasingly astonishing discoveries forced me to remain a whistleblower. We could see the writing on the wall as early as 2005. By 2012, it had become a flashing billboard. At the time of this writing, we have unclothed the emperor and he is naked. We have come to understand the pernicious toxicity of pesticides and the GMO crops developed to incorporate them, and even more importantly, how the lobbies we had to fight were covering it all up.

Chapter 2
DISCOVERY

The genesis of the Monsanto Papers, in which we begin to get a glimpse of the Monsanto's pseudo-science.

Lobster Omelet

January 16, 2016. I was having lunch with the American lawyers for the first plaintiffs against Monsanto at the famed restaurant La Mère Poulard on the Mont Saint-Michel, and the tide was low. The women expressed their wonder at the shifting sands that surround the island. We could hear the rhythm of a whisk beating the eggs in a copper bowl a few tables over: Someone was making one of the restaurant's renowned omelets, in this case with lobster in it, one of the famous specialties of the historic place. My mouth watered at the aromas, my eyes feasted on the sight of all the old stone, but my throat was tight with disgusting anxiety; my research had yielded unsettling results. I already knew I would need all the help I could get from friends to support, write, and publish it.

The American lawyer, Kathryn Forgie, had come to Normandy to ask for my advice and to enjoy some time exploring the area with a colleague. She wanted to gain a better understanding of the independent studies I had performed in my lab at the University of Caen. Not an unusual meeting for a researcher such as myself. How to begin to address the possible connection between Roundup, which contains glyphosate, and its multitude of cancerous users who were claiming overexposure to the pesticide? I took as a starting point our discovery that Roundup contained hidden toxics (besides glyphosate, the better known one) and suggested she start by taking a look at the results of the tests Monsanto had already performed and documented in reports kept under lock and key. What came out of those discussions was nothing less than explosive, and the shock was felt around the world.

In March 2017 the first Monsanto Papers[1] were released. They were a part of close to two and a half-million documents spanning a period from the mid-1970s (glyphosate was patented in 1974) to 2018–2019. Lawyers were bringing out new ones every week relating to different issues. The industrial giant was leaking like a sieve, a situation never before seen in modern times. There were enough lawsuits there to last a century. As Kathryn recalls: "Once the process had begun, lawyers were in their right to question Monsanto's top brass, and request specific documents.

The company handed over millions of documents so that we would drown in them. Not only that, but some of them were flagged as confidential by the judge. Rather than identify the documents that were actually confidential (in relation to intellectual property)[2] Monsanto flagged almost all of them as confidential. The lawyers had to go back to the judge to request the documents be declassified so they could be made public. The judge consented in 2017, and thus, the Monsanto Papers were born.[3] Baum Hedlund, the original law firm involved in the affair, working with other law firms in the United States, took on one sick plaintiff after another. All the work was done contingently, advancing the costs and only getting paid if they won.

Thinking it was being clever, Monsanto had in fact completely opened itself up to legal scrutiny. Little by little, the press began to publish front-page stories about their almost unprecedented corruption of science, of regulatory systems, public health agencies, government officials and academies.[4] Their "rotten-science" is the cause of sickness and death around the globe. I became increasingly nauseated as it became clear that the "experts" paid by the company knew all along about the toxicity of their products, but were time and again duplicitous.

Their number one priority was to falsify the scientific evidence in regulatory documents. Indeed, they hid the deleterious effects of their products in several elaborate ways. It became clear to us that these were adopted by other companies around the world. We exposed several elaborate procedures they used to fabricate their "evidence." They didn't stop there. In various ways, they also permanently destroyed the reputations—even possibly the lives—of researchers who discovered the toxicity. These lies were gobbled up by the press or other underlings, who whipped up fake stories up to and including the "Seralini Affair" itself. This is how I ended up at the center of this unbearable havoc, which continues to this day. It pains me to read the details of the innumerable misdeeds of these so-called "experts." They are responsible for delaying any political decisions that might have banned toxics that daily poison not only more and more humans, but also biodiversity itself.

Our meeting with the lawyers and associates of the Los Angeles[5] law firm Baum Hedlund in November 2018 shows how I became the worst nightmare[6] of one of the largest international manufacturers of GMOs, pesticides, and pharmaceuticals in the world. And it didn't stop there. In June 2019, a member of the French National Center for Scientific Research (CNRS) distributed a sixty-page report, glossy color cover and all, entitled *The Seralini Affair: The Dead-End of an Activist Science*. He had never conducted any research on the topic, but had many opinions on it, nonetheless. His lobby distributed the report to the French Parliament. It focused on the Foundation for Political Innovation's budget; they were busy working on banning Roundup, whose toxicity I had proven. With the release of the Monsanto Papers in the US, juries of the people were awarding hundreds of millions of dollars to punish Monsanto for their duplicity. Meanwhile, in Europe and elsewhere, their masked emissaries kept on disseminating their smear campaigns and advertising the supposed harmless nature of their products as if nothing had happened.

At every level, from the top brass at Monsanto to government officials, from scientists at the most prestigious universities to international regulatory agencies, from publishers of scientific reviews to the media—and often while being duped—people were being bribed, put to work, or used to sell ever more Roundup. We now have proof that the methods they used were worthy of any mafia: the deliberate corruption of the scientific media; the publishing of public health regulations in place and in the guise of regulatory agencies; the keeping of illegal detailed files (even down to their leisure activities) on people who might be of use to the company—people who could be influenced, or who were in a position of power. And let us not forget the ridiculous shenanigans they used to try to retract the most inconvenient study of Roundup and the GMOs that were created to absorb it.

Published for the first time in 2012,[7] "withdrawn for reasons of corruption" in 2013,[8] and republished with all its results detailed in open source in 2014,[9] it remains to this day (2021) the only serious long-term study of the toxicity of a pesticide and its associated GMO. The study has since been corroborated and shored up by other more thorough studies by the reputable scientific editing group of *Science Report* journal published by *Nature*, not only on Roundup but also on the GMOs associated with it.[10] How does Monsanto escape all this scrutiny? Once they put a stop to the publication of prejudicial papers—their experts spend an enormous amount of time on this—or, if one slips through the cracks, they have it retracted for being "compromised," then they act like nothing happened. Public relations firms working on behalf of these multinationals spend inordinate amounts of time on this sort of thing. "Keep moving people, we've got this covered with our drugs—your agriculture, your safe food, your garden, your family, and your health."

The Monsanto Papers and the various court documents could be the makings of a surrealist myth, with, at their center, a malicious plot conjured by a multinational corporation. This myth becomes reality in the Papers: they contain evidence of secret strategy meetings to undermine the credibility of or to dissimulate studies on the toxicity of the company's products. This is especially true for Roundup and the genetically modified seeds produced to absorb it. The Papers also contain numerous letters citing "the Study," as our paper was called. They went so far as to declare in court that no long-term study on Roundup existed.[11]

In the meantime, they played dirty politics aimed at slowing down the research. To this end, they were methodical, and would stop at nothing. They invented simplistic and erroneous scientific arguments with the aim of creating a whole lexicon for use by Monsanto's PR agencies such as the Science Media Center—present on every continent—the media, the scientific community, and politicians. They didn't shy away from using bribery or threats. They ghostwrote for experts under contract who leant their names to Monsanto-sponsored manuscripts and passed for being independent. Among scientists involved in biotechnology and petrochemicals, the source of pesticides, this kind of recruitment is fair game. All this in the name of creating an artificial scientific community that could keep government agencies and scientific journals in line. The members of this group became talking heads for the media and spoke out in the major news outlets despite their undeclared conflicts of interest.

No matter which way you look at it, this strategy is deadly for the common man. I'm going to describe it in detail in light of previously unpublished data and as I witnessed it firsthand and rubbed elbows with all these types of people in the regulatory agencies. I also of course directed the research that provoked all the controversy, and then went on to win seven lawsuits that are only now getting attention (between 2011 and 2017)[12] against Monsanto's lobbies and support networks. The corporation's schemes have calamitous long-term consequences for the health of populations, but they also affect the private lives and careers of the whistleblowers. By attempting to grind them to a halt, they are simultaneously destroying people's lives.

Take as an example the Argentine families whose children are deformed and accuse planes' spraying of Roundup. Or the Sri Lankan tea and sugar cane farmers suffering from deadly kidney disease. And then there is the case of the American groundskeeper Dewayne Johnson who emerged victorious from the first Roundup trial against Monsanto, thus becoming the pioneer for 13,000 further plaintiffs in the consolidated lawsuits Kathryn Forgie started putting together in February 2016. Dewayne Johnson will not live to see his grandchildren grow up: he has a crippling form of skin cancer caused by the vats of Roundup he carried on his back every day . . . to spray on school playgrounds. In early 2019, another patient, Edwin Hardeman, also won a case against the company. And it just keeps on going: In November 2019 the plaintiffs numbered 42,000; in early 2020, the investigative reporter Carey Gillam, together with the Baum law firm, counted over 100,000. These plaintiffs are represented today by many different law firms. Thousands of sick plaintiffs in Canada are now following suit, and in 2019, got together in a class-action lawsuit to demand that the makers of Roundup pay $500 million.[13] There are at least three class action lawsuits waiting to take place.

It doesn't stop there. There are other victims too, the silent ones who have traces of Monsanto's products in their bodies or their urine, whose lives are ruined by terminal illnesses. The fears the public has in regard to Roundup in particular, are nothing compared to the reality. In this book I prove that their fears and pain are far from being irrational, as some like to call it. And it's not just my intuition or that of my coauthor. It is based on my thirty years of groundbreaking research, published in dozens of places, and joined now by thousands of other papers that shed light on the hidden and long-term mechanism behind Monsanto and others' treacherous ongoing manufacture of toxic products.

Neither Bayer, which now owns Monsanto, nor Monsanto themselves have ever acknowledged any wrongdoing. And yet we now have proof that they knew about the toxicity of their products. During the Johnson trial, the members of the jury found that the firm acted with malice or oppression.[14] Judge Suzanne Ramos Bolanos proclaimed in her California courtroom[15] that Monsanto had acted with "malice, oppression, or fraud."

In answer to question number four, "Does either RoundupPro or RangerPro present known or unknown potential risks during their fabrication, distribution, or sales in light of the scientific knowledge generally accepted by the scientific community?" YES, the jury concluded.

As early as 1984, the Environmental Protection Agency (EPA) had signaled the toxicity of glyphosate, but at the time Monsanto pressured them to bury their report.[16] At the end of the 1990s, toxicologist James M. Parry, who worked for Monsanto, never published his report showing the genotoxicity of herbicides that contained glyphosate. It showed that the herbicides reached the first threshold of carcinogenicity. Monsanto stopped him from publishing it and buried it (the report—because the man himself died of a terminal illness). The firm had even tried to get Parry replaced, and to suppress his ideas, according to the Monsanto Papers.[17] And then in 2013, EPA toxicologist, Marion Copley, bid her boss to stop being dishonest and to recognize that the herbicide caused cancer.[18] Since at least 2005 to my knowledge, Monsanto has not let up on the pressure.

In the summer of 2018, Monsanto was fined US $289 million thanks to the law firm Baum Hedlund, Dewayne Johnson's lawyers.

This was right when Bayer was trying to acquire Monsanto. The German giant lost US $17 billion because Monsanto lost the case, and then even more—40 percent of their stock's value—in April 2019. In 2020, under pressure, Bayer settled a case against 100,000 plaintiffs for US $10 billion. Tens of thousands of them turned down the settlement and are fighting onward.

With one piece of evidence after another, we will unravel before you the lies lurking behind Monsanto's public statements and the ones they spread via their network of scientists,[19] scientific publishers, regulatory agencies, PR firms, or via social media, trolls, and reporters. They've known since the start, and they know still, like Bayer and others, but they choose to act in a manner that is deleterious to our health and that of our children. It is a striking example, and a pure and simple illustration of what happens all around us in other areas.

Chapter 3
THE WAY OF THE
WHISTLEBLOWER

In which we pick through the professor's résumé while he steps out for a minute.

Gilles-Eric Seralini has been a professor of molecular biology teaching toxicology and the origins of modern risks since 1990. Today this forms part of the transdisciplinary "Network on Risks, Quality, and Sustainable Environment" that he codirects at the University of Caen.

He reached the level of "exceptional class" in 2017—the highest level you can reach as a full research professor in a State University—no one has published more than he has in terms of scientific research on the topic of the toxicity on mammals (and humans in particular) of GMO crops and pesticides, on their mechanism of action, and our possible detoxification. (His studies that are cited throughout this book can all be downloaded from his website www.seralini.fr.) He has been studying the effects of food pollution on our health since 1984. He has been called upon as an expert by the European Union, for court cases in France and several foreign countries (Germany, India, The Philippines). He is also an advisor to foreign governments and politicians (Canada, Italy, Luxembourg, Austria, The Philippines, Australia, New Zealand, India, and China). In 1998 he was named "scientific expert" by the French Ministry of Agriculture and the Environment, a position he held until 2007. He received the National Order of Merit award for his professional lifetime achievements.

As early as 1991, Seralini was advanced to full professor. This made him one of the youngest university professors in France and paved the way for him to have his own research team, to be able to choose and manage his contracts and research topics, and to achieve autonomy in both his choice of expertise and in what he stood for.

A short synopsis of his career shows where his work came up against multinational corporations, in particular Monsanto, and explains why the latter placed him, more than anyone else, at the center of their preoccupations. They deployed an impressive amount of energy and resources to depict him as your basic activist, when it's clear from just a few lines of his resume that he's no leftist militant nor is he an organic carrot-touting ecologist.

1991: He began teaching about GMOs as chair of a new department of molecular biology at the University of Caen, following four years spent in the foremost labs of North America developing GMOs for research purposes.

1992: Put together an international conference to which he invited a large share of the academics who included molecular biology (the GMOs makers) in their curriculum, as well as senior academics who were at the forefront. At the time, France was one of the world leaders in experimental GMOs as they were thought to reduce the use of pesticides. Seralini included information from the files he obtained in his teaching for students.

1993: The regulatory committees did not want to give him access to the GMO files, because of industrial secrecy.

1994–1997: Seralini obtained findings (syntheses) of the regulatory evaluations of GMOs and analyzed their scientific deficiencies, which he then made part of his teaching and which he talks about in his books and to the public.[1] He saw clearly that GMO crops would increase the use of pesticides rather than decrease it. He posited that breast cancer and certain other illnesses are caused by something in the environment (up to 95 percent of breast cancers are of non-genetic origin). He gave a first conference that took place at the Ministry of Research on the improvements needed in health testing ahead of the arrival of agricultural GMOs on the market. Seralini called for a moratorium before they hit the market, citing the need for further research, and cosigned it with the biologist Jean-Marie Pelt and over one hundred scientists. Some of these scientists were already known to the public in France, such as Francis Hallé, Jacques Testart, Albert Jacquard, and Hubert Reeves.[2]

1998: French regulatory government agencies asked him for critical assessments of the GMO files. The controversy exploded.

1998–2005: Researched the endocrine disruptors (specifically pesticides and plastic compounds such bisphenol A (BPA), and the origins and treatments of breast cancer. He began to speak out more and more forcefully and publicly in opposition to members of the committees he sat on. He rejected the files provided by industry for failing to address health effects, making his votes heard in official reports. Some of his colleagues on commissions didn't lend much attention to the inconsistency of the reports, which were more promotional than scientific (they promised, without data, that GMOs will feed the world, be good for the environment, and reduce the use of pesticides). Some of his colleagues were openly critical to Seralini's positions, and they began to vilify him. The gauntlet had been thrown. He would face them in court during the defamation cases he filed—and won—against Monsanto's lobbyists between 2011 and 2017.

2004–2005: Gilles-Éric Seralini proved the toxicity of Roundup on human cells in his lab. The product was used on 80 percent of agricultural GMOs. The glyphosate it contains was confirmed to be harmless on its own. After the study appeared in 2005 in the major journal *Environmental Health Perspectives*, the editors came under such pressure that they asked Professor Seralini not to publish with them anymore. We now know, thanks to the Monsanto Papers, that the company criticized the article in order to prevent its publication.

2005: During his third term on the French GMO regulatory commission, Professor Seralini finally received word from Monsanto that the company had completed the first three-month tests that were requested seven years earlier (for a corn that produced an insecticide, MON 863), but it proved impossible to get ahold of the data. Yet this was the mission he had been assigned, mandated by the government even though he wasn't paid for it. He knocked on every door: the French government, the commissions he sat on as an independent expert, the European Food Safety Authority (EFSA), and the European Commission. In the end he received the data from the German government, prompting Monsanto's ire. The company went to the court of appeals to try to keep the results confidential but lost.[3]

2006: Monsanto's Bruce Hammond published a paper assuring that the GMO had no health repercussions in a journal that welcomed the company's contributions, *Food and Chemical Toxicology*. But this author did not declare any conflict of interest nor provide any raw data. Professor Seralini published a counter-study in another scientific journal[4] based on the documents finally obtained by the German court, and omitted from Hammond's report, showing the first signs of hepatic and renal toxicity. Monsanto reacted a few months later.

For the first time in the history of science, a living researcher's name was included in the title of the article. It appeared in the same *Food and Chemical Toxicology* and pointed a finger directly at Seralini. The journal belongs to Elsevier, one of the largest publishing conglomerates, with an annual turnover of over 7 billion Euros. The experts on Monsanto's payroll were such fixtures of the journal that they paid off its editor in chief in order that they may dictate their own rules, including the ability to have Seralini's main article retracted in 2012 soon after it was published. (We'll come back to this.) **2009:** Professor Seralini continued to publish his research, accomplished with the help of his Canadian colleague Louise Vandelac, on the risks of transgenic salmon. Their report blocked the authorization to commercialize the salmon. After ten years of intense lobbying, the transgenic salmon case is not so fresh anymore, but soon we'll be able to enjoy this Frankenfish—as they say—in North America; it will be imported thanks to the transatlantic Comprehensive Economic and Trade Agreement (CETA). They won't give up!

In 2009, Seralini also handed in a report to the Quebec government on hormonal disruptions caused by glyphosate-based herbicides. But most importantly, he published his joint research with experts the world over who denounced the stupidity[5] of the arguments Monsanto resorted to when they attested to finding nothing to raise any concerns in their test results. He continued to believe it was possible to move the regulation forward based on the risk assessments by exponentially increasing exposure of the evidence in books, in international conferences, and in scientific journals.

At the same time, he studied the data on three GMOs tested on animals by Monsanto over only three months. He obtained these thanks to the German court order of 2005. This is how he came to show the signs of toxicity on the kidneys and liver of the two main types of agricultural GMOs. But above all he exposed Monsanto's lackadaisical behavior, that was covered up by the regulatory authorities in France led by a certain Marc Fellous.

2010: Following a documentary on TV France 5 about Professor Seralini's research, the very same Marc Fellous (who, although he was a specialist of the Y chromosome for the Institut Pasteur, had never published any research on GMOs) sent a letter to the *Conseil supérieur de l'audiovisuel*, a regulatory agency for all things audiovisual, entitled "*Le Magazine de la santé*: une émission militante?" questioning whether the show was militant. It was laden with defamatory remarks about Dr. Seralini, who filed a lawsuit not only against Fellous, but also the organization he belonged to—the *Association française des biotechnologies végétales* (AFBV)—headed by Axel Kahn, a renowned geneticist, and that included members of the *French Academy of Science and Agriculture* who granted permits for GMOs. The study's detractors all had ties to Monsanto, as it will be proven in the investigations into the two lawsuits, and later, in the Monsanto Papers.

In his laboratory, Dr. Seralini put in place and finalized the protocol for a study on the life-span of the lab rat (two years) to determine if GMOs and the Roundup associated with them are toxic or not in the long run.

His work allowed him to understand how human cells become poisoned by pesticides and pollutants as early as the fetal stage. This led him to direct some of the members of his research team to examine possible ways to detoxify the pollutants. Unbeknownst to him, this work would bring him head-to-head with Monsanto's lobbyists. The Monsanto Papers show how one of Monsanto's toxicologists, William F. Heydens, was congratulated by Monsanto for getting picked as a proofreader for the journal *Regulatory Toxicology and Pharmacology*, and as such for blocking the publication of the Seralini team's first discovery on the topic in 2011 (MONGLY03081997).

[Monsanto document reference, discussed in more detail in chapter seven.]

He attained notoriety for his *cri du coeur* (outcry), which revealed all the hatred which would crystallize further: "Seralini is a persistent bugger" (MONGLY02722492). This is when it became apparent that the journal was actually an organization funded by the industry with the aim of deregulating its products, such as tobacco. A certain Henry Miller had already played an important role in authorizing carcinogenic additives to tobacco. This was all denounced by scientists and even reached the US Congress, who investigated him for having also, for example, influenced the Food and Drug Administration during the approval process for an endocrine disruptor, Bisphenol A. If it isn't one poison, it's another in their practices.

During a seminar,[6] Professor Seralini commented on the incident. "Even as I encouraged my students to try harder (they needed to in order to pass their exams) I felt the blow of the rejection of my study's publication. It was important research among other things on the body's ability to detoxify. Not quite understanding the reason why, my team and I were in shock. It was almost laughable. The reasons given by the editors seemed bogus, asking us the same questions we had answered three sentences earlier. I felt like I was interacting with a secret society that would use every trick in the bag to reject us rather than point out to us the possible flaws in our study—which would have been their role under normal circumstances. What did they have against the detoxification of Roundup brought about by plant aromas? Didn't they want people to get better? I was highly annoyed. I would never have guessed that a Monsanto toxicologist could have held so much sway over the journal's biochemistry articles. It has all become clear." Dr. Seralini published his study elsewhere.

2011: On January 14, Monsanto went so far as to enlist the help of a competing firm, Syngenta, to fight the publication of Seralini's study demonstrating the toxicity of Roundup (MONGLY03081997). It seemed they have more interest in teaming up against him than in competing for pesticide and GMO market share.

(Author's Note: Moreover, reviewing the papers without indicating their links to Monsanto, confidentially in the journal, was a fraudulent practice.)

2012: On September 19, after three years assembling the budget, one year obtaining and growing the crops, making a detailed protocol to study the animals, two years of experiments on the rats, one year of calculating, coming up with statistics, and writing, four months of analysis and second opinions from scientists around the world hired by the journal to fact check, the famous study was finally published in *FoodChemTox*, provoking "the Seralini Affair." The part about GMOs made headlines around the world, but they failed to mention the toxicity of Roundup itself, possibly leaving out the most important part. The photos of the rats' tumors did, however, make the front page—images that so terrify Monsanto.

As Seralini said during a conference: "I was at the end of my rope, thinking I was nearing the stage in my work that would logically lead to the ban of GMOs and glyphosate-based pesticides—that we could turn the page. As the publisher and the scientific reviewers themselves pointed out, it seemed as if the scientific community should have acknowledged that this was the most thorough and extensive study ever completed on lab animals to demonstrate the effects of a pesticide—the most commonly used one in the world— at a very low dose, and of the GMO created to absorb it. I was shocked by the meanness of the instantaneous discourse in the press by the so-called independent experts who turned out to be the very same people who had authorized the product's use. Just when I thought I had come out of the throes of a huge amount of work, I was in fact getting hammered by a cyclone of hatred, relayed by the incessant repetition of the same turns of phrase and preposterous arguments the world over. They made any measured or serious debate impossible. At the time I couldn't call the firm out on their corrupt entanglement with the intricate network that we know about today."

From 2012 on, Dr. Seralini's research continued, showing the mechanics of toxicity in further detail. It's worth pausing for a moment to recognize just how explosive things were at this particular point in time: "Their counter-arguments were so rude and violent that I was determined to dig further; they were probably hiding something worse."

Chapter 4
THE EXPLOSION OF 2012

In which the media, the politicians, and the scientific and regulatory intelligentsia emit a cacophony of annoyance and get angry at me, but we still don't know who's in charge.

Way before *Forbes* magazine denounced Monsanto as Miller's ghostwriter, the company published the two incendiary pages[1] that the international press seized upon and that cemented the notion that Seralini's 2012 study was "fraudulent." The original two-page text, written and revised by Monsanto, was kept confidential and was tucked away in the emails of the company's experts. Monsanto even had to tone down Miller's vitriolic statements about me *via* Eric Sachs. The piece was handed over to public relations firms such as the Science Media Centre in London and others, reaching as far away as Australia, New Zealand, and France. Scientists and journalists seized upon it as well, though following the lawsuits I filed against them they were later found guilty of libel.

Let's stop for a moment to talk about the Science Media Center (SMC). The SMC is a case in point of the lengths industry will go to in order to influence the media—and hence politicians—and to get the authorities to deregulate and provide other favorable conditions for corporations such as Monsanto to maximize profits. Established in London in 2002, the SMC is hosted by the Royal Institute and "is delighted to have close ties with the British government."[2] It is science's equivalent of a Trojan horse, financed by pharmaceutical companies from around the world seeking to safeguard the public's ignorance of their misdemeanors. Its supporters include the cream of the crop of the oil and pharmaceutical companies, the food industry, agrobusiness, as well as retailers, united in lobbying entities such as the International Life Sciences Institute (ILSI), and also, more surprisingly, in two of the most prestigious scientific publishing groups—Nature and Elsevier!

Over twenty similar centers that produce "industrial science" (in a united voice) exist around the world—from Japan, to Australia, New Zealand, and the United States. According to Tom Wakeford, a biologist and reporter for the *Guardian*, the SMC has become "the prime minister of the Truth, and something the fiction writer George Orwell would have been proud of."[3]

Science reporters refer to them for ready-made copy written by ghostwriters that they can quote verbatim, reports by experts hired by industry, and conveniently apropos scientific papers. Many science reporters, or those pretending to be, do no more than lift quotes from the SMC without even bothering to check other sources. This is how, when Seralini's study came out in 2012, two of the six articles that appeared in *Forbes* to try to topple the research on the effects of Roundup and GMOs were direct quotes from the SMC.

Everywhere I went, from Australia to Japan, passing through India, the Philippines, and Canada, my arrival was preceded by vehement articles in the press and smear campaigns against my work or my person, carefully orchestrated by the SMC and its offshoots. The words of Henry Miller, picked up by the SMC, were thus rebroadcast by news agencies in every country where I was invited to attend international conferences, and usually just before they took place. Bloggers, hiding behind pseudonyms, referred to the same information to rewrite my Wikipedia page weekly to their heart's content, or to cook up the same story in all languages that they called "The Seralini Affair." This is the shortcoming of the "living dictionary" whose sources are often subjective and by unknown authors, a particularly vexing quality when it comes to controversial facts that have an important economic impact.

Indeed, my 2012 study is one of the most consequential in modern science and one of the most televised (the images of rats with large tumors). As a leading sociologist at the School of Advanced Studies in the Social Sciences (EHESS)[4] proclaimed in 2012, two months after the original study was published: "There will be a before Seralini and an after Seralini." It was qualified as a "shocker," and it provoked a "worldwide outcry."[5] Even in a critical review, a journalist from the *International Journal of Medicine* admitted: "Because of the Seralini Affair [. . .] GMOs were almost completely banned in France and in Europe in general."[6] There certainly was a lot for Monsanto to fight for!

When my ground-breaking 2012 study proved the effects on the lifespan of living mammals, of not only Roundup but of GMOs themselves, Monsanto went into overdrive to make it go away. Roundup had never been tested over a long time frame. Glyphosate alone had been tested this way, but even though it is the only active ingredient listed on the label, it represents a mere 40 percent of the ingredients in Roundup.

What set my study apart was its duration in time and number of parameters measured on rats. We conducted blood tests, urine tests, behavioral, and histological tests (over 100,000 in all on 200 rats). We also used very low regulatory levels of Roundup over two years (the life of a laboratory rat). In some groups the rats were fed with the same level of GMOs present in the American diet. Deadly renal and hepatic diseases killed the animals by their second year, even before their large tumors became cancerous. In comparison, the groups of rats fed pesticide-free and GMO-free food did not come down with chronic illnesses, and had between two and five times fewer non-cancerous tumors—and only very late in their lifespan. Up till then, the tests Monsanto had done that led to the commercialization of GMOs only spanned a three-month period in the lives of eighty rats (in these tests 320 rats out of 400 were controls) and under our cross-examination we found that the animals were already in fact showing the first signs of renal and hepatic toxicity. This is why we followed up their study and enlarged it using the same protocols and the same type of rats.

Only the GMOs (half of the study) were included in the subsequent debates and reviews, not Roundup. And instead of using the name Roundup, they would only mention the word "glyphosate." It worked—and still works—for whoever wanted to hear it, from the public to the NGOs and the press. They wanted it retracted so they declared it fraudulent and defamed my person. It was all deceit and confusion, two commonly used weapons.

For if the deceit was brought to light—declaring that glyphosate was the only active ingredient in Roundup—then the company would have to close shop. It took my team thirteen years to unveil and publish the exact nature of the toxic ingredients that Monsanto added to glyphosate in the formulations of commercially available Roundup.[7] The Monsanto Papers reveal that the company's experts already knew the nature of these formulants and their dangers at a much earlier date. Normally a false declaration during the approval process of a product incurs its immediate ban, without recourse to political or scientific consideration. And if it is shown that the same is true for most pesticide declarations, that don't focus on the most toxic formulants used but on the ingredients the manufacturers are willing to declare, then the agrochemical industry itself had no more basis in honest regulation. The stakes couldn't have been higher.

A year after our publication, and after sustaining many more transgressions, we came to understand the singular modern-day inquisition being used that led to the retraction of our 2012 study. Many reporters or distant observers of the polemic let it go at that. And yet, several pointed studies with very precise methods[8] conducted between 2012 and 2018 confirmed the specificity of our results.

Chapter 5
IT'S TRUE BECAUSE WE SAY SO

In which Monsanto's lawyers wrangle with the highest court of the state of California to prevent the photos of rat tumors from Seralini's study to be shown to the jury of the people out of concern they will "ignite the passion" of the jury against the firm, and in which we discover the role of GAFAM.

Just before the groundskeeper Dewayne Johnson's historic lawsuit, Monsanto's thirteen lawyers tried on two separate occasions—May 24 and June 12, 2018—to persuade the court to exclude from the deliberations "any mention, argument or reference to "The Study,"[1] the book on the subject of Dr. Seralini,[2] the documentary,[3] and any fact or image contained in any of them."[4] The lawyers for the plaintiff, meanwhile, were preparing to use these.

In another day and age, they would have burned me, my work, and my coauthors in a public space! Monsanto's lawyers kept repeating that "the Study" was flawed and not to be trusted according to the international scientific community they had assembled. In the argument for the defense, the study was summed up as "universally rejected" even though by that time it had been republished. The firm had managed to get it retracted under the "pseudo-scientific pressures"[5]—something they hid—after it had been reaccepted, as we saw, in another scientific journal, but that didn't count according to them. *It's true because we say so,* so goes their increasingly absurd syllogism, *and the proof is that we can assure you.* They repeated ad nauseum in writing that the study was "highly prejudicial" to the Monsanto company, and they declared that the rats "covered in tumors" were dead in the banned photos when in fact they were alive. They went on to assert that "there was nothing to discuss" [about the possibility that Roundup causes cancer]. Yet this was the crux of the court case. They emphasized this in their obstructive way, denying everything in one swoop. As they put it: "The reason that *top scientists*—the ones Monsanto went to great lengths to recruit, as revealed in the Monsanto Papers—rejected the study could be partly because the images [of the rats] were shocking to the public and their conscience."[6] American law allows images deemed too shocking for the jury to be excluded from a case. Facebook also refused to allow the photos to be posted, in 2019 and 2020. Meanwhile the lesions on Dewayne Johnson's skin didn't seem to shock the firm's lawyers one bit.

Another multinational in the GAFAM (Google, Amazon, Facebook, Apple, Microsoft) network, Google, also participated in its own way in the disinformation of the public thanks to its location-based advertising policy known as *geofencing*. This is how over a holiday weekend during the Hardeman trial in San Francisco, ads on Google appeared on local smartphones and computers, promoting the safety of Roundup. One website in particular, *Weeding Wisely*, appeared first in various Google searches. It featured articles such as "The fear of chemical agents is based on a misunderstanding," or "The media onslaught against weed killers lacks scientific support," or even "Pay attention to the science behind glyphosate-based herbicides [in other words Monsanto's] not to those who want to scare you." (Like us, with the tumors!) Even Judge Vince Chhabria, who had given the lawyers for the plaintiff a hard time at the beginning of the trial, recognized the extent of the phenomenon and warned the jury not to be influenced by such off base tactics, but only on the evidence presented in the trial itself.[7]

Beyond Monsanto's pretense of wanting to spare the jury the sorry sight of helpless animals deformed by its very own products, it was a question of keeping the most important study on the effects of the Roundup brand from consideration, substituting it in the deliberations for a decoy—glyphosate.

Chapter 6
THE MAKING OF A LIE[1]

In which a gangster flees in his car to protect a multinational corporation that's going all out to keep secret a huge lie that underlies the toxicity of the agrochemical industry. In which we see the firm get cornered in lawsuits, and in which we discover the chemical composition of the many Roundup products.

A beautiful red sedan took off with a bang and blew through a stop sign in this well-heeled suburb of San Francisco, as its driver was trying to evade the questions of a French journalist.[2] Tristan Waleckx, of the popular French TV show *Envoyé spécial* (Correspondent) was trying to get an interview with Henry Miller. He was the one the US media even nicknamed "GMO *cheerleader*." He was already well-known for his attempt to reintroduce DDT and for his defense of pesticides in general. He was also known for siding with the tobacco industry and their toxic additives during the major lawsuits against Philip Morris. He joined the campaign against organic farming, widely proclaimed his denial of climate change, and even extolled the benefits to human populations of the Fukushima nuclear disaster.

This scientist from Stanford University worked for the National Institute of Health (NIH) and the Food and Drug Administration (FDA) up until 1994. This was how he nurtured his ties to industry. During those years he published in major biotechnology journals on the possibility of modifying human sex cells. In 2018 he published on the need for transgenic flowers. He was a real jack of all trades who promoted the interests of industry, even when it came to the GMO vaccine against the flu. He and his colleague Bruce Chassy had both worked for Monsanto at various times.

He was being chased by Tristan Waleckx because he didn't want to answer questions about his articles that were ghostwritten by Monsanto, omitted from his CV. In 2015 for example, a higher-up at Monsanto, Eric Sachs,[3] who considered Miller to be a "magic pen," handed him the rough draft of an article on the harmless nature of glyphosate in response to the IARC[4] study that recognized the substance as a probable carcinogen. Miller lifted 85 percent of the piece, changed a comma here, a few synonyms or a grammatical construction there, and published it under his name in *Forbes*.

Not until 2017 did he get dropped by the magazine, and his contributions erased, when the truth about the ghostwriting came out in the Monsanto Papers. But he had already accomplished his goal, of spreading his propaganda, and now he was taking one for the team. Unfortunately, all of his articles, similarly ghostwritten, inspired journalists the world over who didn't pay attention to their sources.

In 2013, after a year-long investigation at Monsanto's request, the journal that had originally published my study[5] had to admit that there was no evidence of fraud.[6] But the allegations in the media persisted. These defamations once again insinuated that I was at the very least controversial, and as such, I must be excluded from the debate. It was unbearable. We filed lawsuits against the three most highly visible cases of defamation. These same journalists insulted mayors of towns where there was a ban against the use of pesticides near houses with children living in them. Meanwhile, recognition by juries of the people that there was a link between the use of Roundup and cancer had Monsanto and the entire community of pesticide producers, agencies, and the press writhing in discomfort. What's more, the firm was penalized for its dishonest evaluations in which it demonstrated it was much more interested in the reputation of its products than their safety, according to the judges.[7]

The lawsuits kept coming. The one filed by Alva and Alberta Pilliod—a husband and wife suffering from cancer caused by Roundup—began on April 1, 2019, in the wake of Dewayne Johnson and Edwin Hardeman's wins against Monsanto-Bayer. The jury verdict on May 13 marked the dramatic beginning of the end of the Monsanto Years, as Neil Young put it in his song. The famous Canadian musician is quoted tens of thousands of times in the Monsanto Papers as a vehement critic of the firm, particularly in his album *The Monsanto Years*.[8] In fact, we learn from the Papers that Monsanto did an analysis of his lyrics to determine whether Monsanto was mentioned and what damage this had on Monsanto's reputation; and even sent spies to Neil's concerts to see what effect he had and what he was saying about Monsanto.

As an aside, when Leemon McHenry sent the Papers he discovered to Michael Baum and Robert F. Kennedy Jr. at the first trial in San Francisco, Robert called Neil Young and invited him to attend the trial. Neil showed up the next day, much to the annoyance of Monsanto's lawyers. His appearance created a media sensation. Michael showed Neil the documents, including the one where they follow him around and spy on him and another where they tell Monsanto employees that they must not admire his music any longer (or something to that effect). Neil told Michael that he greatly admired what the lawyers were doing, and it made him want to return to college. His wife reminded him he never went to college in the first place. Michael then told him he deserved a PhD for writing all those great songs.

The point is, the Papers proved the firm's bad faith, helping victims to win further lawsuits. The overwhelming evidence put an end to Monsanto's strategy of calling out critics and stripping them of their credibility with fake science. The Pilliod trial proved more portentous than previous ones. Indeed, Monsanto-Bayer was accused of "ill intent, pressure, and fraud committed by one or more of its agents, directors, or managers all acting in the company's interest." The next question was when they would condemn those responsible. The judges were struck by the group's stubborn refusal to accept the evidence, to recognize its wrong-doing, and to negotiate with the victims, and they become hardened. Even as French TV continued to boast the virtues of Roundup, the jury and the judge went along with the lawyer for the plaintiff, Brent Wisner, and his recriminations, slapping Monsanto with a punitive US $2 billion settlement for the Pilliods.

This represented an exponential jump in scale compared to the previous cases, and that astounding sum was added to the US $55 million granted to the couple for the lymphoma they developed as a consequence of using Roundup. According to Brent Wisner, "this sum represented two times Monsanto's annual profits, elegantly rounded up to one billion dollars each." After the Pilliods, 100,000 plaintiffs awaited their turn. The edifying verdict forced the firm to negotiate with a number of them. This would be the third lawsuit for Monsanto as they continued to deny their wrongdoings. They paid the damages, more preoccupied with propagating false science than actually studying the toxicity of their product, that many continue to call glyphosate, rather than Roundup.

As we mentioned earlier, Bayer had already recently lost 40 percent of its value. This only got worse, to the tune of 37 billion Euros lost since the first lawsuit. In the summer of 2019, they appealed and managed to lower the damages, in keeping with the American judicial process. In the meantime, Monsanto resumed marketing its glyphosate-free version of Roundup. The big question was whether this also contained hidden poisons.

The lawyers kept on coming up with documentary evidence. If at first Monsanto fought against the 2012 study, they now had to maneuver and sidestep over twenty papers my group had written that came up in the deposition. Indeed, those studies,[9] even the latest ones, often highlighted the difference in toxicity between Roundup and glyphosate. Glyphosate was by now well-known as it was all the media and the regulatory agencies talked about. It is the only toxic substance the company lists as an active ingredient in Roundup.

Taken alone, glyphosate didn't appear to be that toxic in the eyes of the regulatory agencies. The acceptable daily intake, or ADI, is the daily dose that theoretically doesn't pose a health risk. It was calculated solely on the basis of the listed active ingredient, glyphosate, without taking into account all the hidden poisons found in the rest of the formulation for Roundup, the product sold to the public.

Little reminder: The notion of an ADI was put in place back in 1961 by the Food and Agriculture Organization of the United Nations and by the WHO,[10] at industry's suggestion, with recognizable spokespeople such as René Truhaut in France and Europe. "We had to find a legal framework for the dissemination of thousands of different chemical molecules in food. [. . .] Without an ADI, the system wouldn't work. Thanks to this grandiose anesthetic for the public, human beings can believe they are protected by an army of valiant and disinterested scientists," wrote Fabrice Nicolino.[11]

Our studies showed that the herbicides that contained glyphosate, such as Roundup, introduced to the market in 1974 and sold under many different brands all over the world, contained, in addition to glyphosate (about 40 percent of the whole), numerous hidden poisons such as half-combusted oil and heavy metals, including arsenic. This explained the many illnesses that Monsanto was determined to keep hidden. The whole world would question the rationale behind the complaints, even as they continued to focus on glyphosate alone. This is a crucial if not well-known fact.

The company managed to blow enough smoke in the public's eyes that not one reporter mentioned the cover-up, even the ones seeking a higher degree of transparency, such as the correspondent for the TV show *Envoyé special* (Correspondent).

So why were there so many different illnesses? Dewayne Johnson got skin cancer. He carried dripping cans of Roundup on his back as he sprayed hundreds of liters of it per day on schoolyards. After prolonged contact with the skin and the mucous membranes, arsenic will cause skin cancer at the doses found in Roundup products. But the exchange between the International Agency for Research on Cancer (IARC) and Monsanto focused solely on glyphosate as it was the only active ingredient in Roundup that the company declared to be toxic. A major fraud. As it turns out, Monsanto began their defamation campaign as early as 2005,[12] when my team member Sophie Richard demonstrated the difference in toxicity on human cells between glyphosate and Roundup.[13]

That is the key problem: Monsanto focuses the debate on glyphosate alone.

In 2019 glyphosate was banned by various governments under pressure from the public, but only for individual and not for agricultural use in France and a few other European countries. This is why Monsanto began selling glyphosate-free Roundup. All the NGOs who had cried out and petitioned against glyphosate rejoiced. All the while advertising for glyphosate-free Roundup made a comeback with a vengeance. Monsanto's dog Rex, who was retired after a false advertising campaign on the biodegradability of Roundup years earlier, made a comeback on TV. Rex returned as Monsanto's familiar good old dog of Evergreen Garden Care, the distributor of Bayer Garden. But he seemed to have changed owners. Indeed, in 1954, Monsanto and Bayer had already partnered under the name Mobay, and then later on again in order to sell 75 million liters of the chemical weapon known as agent orange to the US government, mired in the Vietnam War. The good old pup roams happily around his garden, this time promoting the merits of the new Roundup that boasts it can kill weeds mercilessly in one hour—and without glyphosate, if you please!

The unwanted weeds that you couldn't get rid of before this product could now be eliminated thanks to Bayer (or its distributors) or with similar products made by its competitors. The new products, according to the labels, contained either vinegar (acetic acid) or a synthetic extract of geranium (pelargonic acid) or even a residue of palm oil (caprylic acid). After being touted as biodegradable, Roundup simply became "organic" in the new ad campaigns. Was the soon-forgotten miraculous molecule of no use? It was, but only to patent the product. Did the hidden poisons change? Not one bit. In 2020, we performed research on the "new" herbicides and showed that they contained arsenic and petroleum derivatives that are still not listed on the label.[14] Sometimes we even found traces of glyphosate. Rex should have been banned from appearing on TV ever again. But they never gave up; only the legal system could make them bend.

Edwin Hardeman developed non-Hodgkin lymphoma, or cancer of the lymphatic system. So did Alva and Alberta Pilliod, the plaintiffs of the April 2019 court case. A few studies demonstrated that at high doses glyphosate causes cancer. As long as the public health agencies continued to debate the issue, Monsanto still had a chance. And yet, the link between those lymphomas and the petroleum derivatives and pesticides that contain them was well established in the scientific literature,[15] it just hadn't entered the debate. The substances in question had to enter the body bit by bit to lead to that kind of illness. That's why it was essential for Monsanto-Bayer to hide the presence of these undeclared substances, and to focus on glyphosate alone. The media complied.

By the way, the chronic ingestion of arsenic also causes kidney failure—a well-known fact—as do petroleum byproducts. The organs responsible for cleansing the body (the kidneys and the liver) were the first ones affected. In Sri-Lanka, farmers suffered above all from kidney failure; they consumed water contaminated with the Roundup they used. The rats in my study that were given the pesticide in its full formulation showed the same symptoms.

We're now going to delve deep into the heart of the Monsanto Papers. If you'd like a taste of it consider this. Dozens of employees or others associated with the firm were working either on the LNG program (Let Nothing Go) or on its sidekick WAM (Whack a Mole). You can think of the moles as the whistleblowing scientists who discover toxicities: If you hit them hard enough, they—or the politicians and supporters who listen to them—stayed underground. To finish them off, Monsanto used defamation or even manipulation. The Monsanto Papers and the court cases that refer to them, shed light on all of this.

Chapter 7
RETRACTION BY CORRUPTION[1]

In which we open the beast's ribcage and gradually expose the heart of the Monsanto Papers, explain what a MONGLY is, and present a chronological look at the fraud that led to the retraction of my disturbing study.

I had now spent five years working and intensively researching in order to put the protocol for my study into place. It was August 2, 2012. I was finally taking a few days off to be with my family in Nice. The cry of the seagulls and the air out on the balcony had rarely had such a calming effect. The European editor of a very reputed international scientific journal, *Food and Chemical Toxicology*, had just confirmed the acceptance of the final manuscript of my study. This followed four months of intense and exceptional (due to the numerous parameters studied) peer-review in which I had to answer questions from experts who make the final decision about whether to publish the article or not—a standard process that works well.

The European editor in question explained that this process was perfect for our paper but lost his job following Monsanto's shenanigans, but first the journal accepted this singular work. The publication date was set for September 19. All the while, manuscripts from every continent in the world were making their way to the journal's American headquarters in preparation for the fall edition. At this point in the process the studies to be published were not yet public, and yet mine was somehow leaked to Monsanto. As early as August 12, two high-ranking employees confirmed their intention to pay Wallace Hayes, the American editor in chief, a consultant's fee to organize a meeting. Wallace was supposed to be independent and under a confidentiality agreement with the journal. Seeing how quickly they struck a deal suggested that they must have already known each other.

A Timetable of Corruption

You may want to take notes. This is one of the best-documented examples of corruption in modern science. You could also tear out the pages of this book and sleep with them under your pillow, so you don't lose the thread of this humanitarian thriller, in which the assassins come up with convoluted ways to avoid being unmasked. Ian Fleming or John LeCarré wouldn't have turned their noses up at this case. It has it all: espionage, secret files, Bulgarian umbrellas, or worse. All this is put into play by a group of dishonest people who are out to conquer the world—something they shamelessly broadcast from time to time via their PR firms. Corruption blows through Monsanto's headquarters in Saint Louis, in Crève Coeur to be exact (which means "broken heart"—you can't make this stuff up), Missouri. It seeps into public and scientific institutions in the form of lobbies. It has a grip on universities, on top researchers, and on politicians. It protects the petrochemical industry and its infiltration of the food supply to the detriment of our health. It allows the regulatory agencies to pump us full of unhealthy products.

We provide a reference number for each of the emails found in Monsanto's millions of secret documents. Each number is preceded by MONGLY (for Monsanto-Glyphosate), which identifies a Monsanto document produced in the litigation.[2] As we saw earlier, the Baum law firm had access to these documents, the most incriminating of which became exhibits that helped win the court cases. They have been released to the public.

To summarize:

1. Monsanto is questioned over the toxicity of Roundup and is forced to share its relevant internal documents with the lawyers for the plaintiff.
2. In order to sow confusion and render the lawyers' work more difficult, the firm remits millions of pages of documents to Baum Hedlund.
3. Narrowing down the papers by key-word search, the lawyers ask the judge to make public the documents they need to prove are pertinent in the court cases to come.
4. The judge can decide whether to release the said documents (only a few at a time). Monsanto redacts personal information in the name of privacy.
5. These documents are now accessible to all and are available on the law firm's website, as well as on that of the Drug Industry Document Archive (DIDA) at the University of California in San Francisco and US Right to Know.
6. Meanwhile, lawyers for the plaintiff are able to call upon the various persons identified in the documents and have these persons testify under oath.
7. Thanks to this trove of documents the court cases were able to prove Monsanto's bad faith. Nonetheless, priority was accorded to the most sickly patients who could prove they had used Roundup.
8. In 2020 the lawyers began to negotiate with the company. They wanted to avoid individual lawsuits and were inclined to cooperate with the class action lawsuits and pay out sums all at once according to the judges' recommendations.
9. In the legal battle against Monsanto (now Bayer) the lawyers did not want the case to lead to the company's

bankruptcy as that would have meant all hope of compensation for their clients (or themselves) would be lost. This point is crucial to understanding the negotiations—that Monsanto could always refuse or delay, until they found themselves back in court.

A methodical chronology will allow us to reconstitute and exemplify, in a precise way, exactly how the gigantic group manipulated science and substituted their own story, or rather, storytelling, in its place. In this case, the myth of innocuity whereby Roundup's main active ingredient is glyphosate alone, and further, that if agriculture didn't use it, famines would ensue. Also, that glyphosate is less harmful than the fat on meat[3] and that you could even drink it by the glass. And yet, all those who made these claims refused to do so in front of the cameras.[4]

These are the events of 2012.

August 2. Final acceptance of the article "Long Term Toxicity of a Roundup Herbicide and a Roundup-Tolerant Genetically Modified Maize" to appear on September 19. The article made its way to the desk of the editor in chief, Wallace Hayes, at the American headquarters.

August 12. (MONGLY00971543). David Saltmiras, Monsanto's manager of toxicology in "new chemistry" (and all combinations thereof), is a regular ghostwriter used to putting the company's communications and announcements under other people's names (MONGLY02145917, MONGLY02145930). He wrote to Bill Heydens, Monsanto's "Product Safety Assessment Strategy Lead," who is also an experienced ghostwriter, published in scientific papers and in magazines, who lived in Henry Miller's elegant neighborhood. He sent him an email asking him, among other things, what the consultation fee would be for a meeting with Wallace Hayes.

September 7. (MONGLY02185742). A contract was drawn up with Wallace Hayes for a fee of $400 an hour, for a total of $16,000.00. As these things go, this was probably just part one. Or just the part that was put in writing.

September 19. Monsanto's interference came too late and the article was published. Scientific journals have strict guidelines. (MONGLY01096619): Hayes wrote to Saltmiras who promptly informed a number of Monsanto employees, in particular Heydens; Daniel Goldstein, pediatrician and clinical toxicologist; and Bruce Hammond, who was responsible for assessing the Roundup-tolerant GMO known by the code NK603 (but he only tested it over three months, whereas we tested it over the lifetime of the animals). Saltmiras also reached out to external contacts and asked them to recruit more collaborators. These people knew me and had already intervened against my previous works that didn't suit them. They were angry. They engaged in an exchange of emails seeking mutual reassurance and encouragement, ways out and strategies, having been without recourse up until this point.

The rat photos had gone viral and certain countries had started to ban the GMO or to question their local authorities. Saltmiras, at a loss, wrote the following to Monsanto's directors: "I suspect that this article could be of use to us, by giving us the opportunity to strip Seralini of the last shreds of his scientific credibility" (MONGLY01096620). He labeled the article "junk science" and quickly recalled his troops. What was extraordinary about these first reactions is that not one of the experts questioned the safety of their product, when they themselves had never run these kinds of tests, and admitted it. They didn't have any real scientific arguments. This too was evidence of their overwhelming guilt. If they hadn't understood the true nefariousness of the product, their first reaction wouldn't have been this one. As it happens, they went looking elsewhere for the grounds for the take-down, gradually bringing more so-called scientists on board.

September 20. The big guns, the would-be experts, like Gérard Pascal, cited in *Le Monde* and other newspapers, played arbiter without revealing whose side they were on, and took a stand without the benefit of any scientific perspective on the media strategy of the Science Media Centre, at the ready "with its sound bites and bountiful quotes from anti-ecologist hitmen" (wrote Benjamin Sourice in his blog on *Mediapart*). Gérard Pascal, the toxicologist who had opposed my expertise when I was on the *Commission du genie biomoléculaire*, or biomolecular engineering commission (who authorized the GMO, under the presidency of Pascal's ally Marc Fellous), "since retired," will nonetheless "make use of his experience by becoming a consultant for the food industry and companies such as Danone and Nestlé."[5] He had been a member of the International Life Sciences Institute (ILSI) since 2010, "the most powerful food lobby, with four hundred members,"[6] among them Monsanto, Total, Unilever, a consortium in which Hoechst and Rhone-Poulenc are in it up to their elbows, as well as Syngenta, Bayer, Danone, Nestlé . . . Monsanto presided over this gigantic association from time to time, among other corporations, with the goal of deregulating the biotechnology and tobacco industries, or denying the toxicity of petroleum residues.

Then in came François Gendre, the director of life sciences at Danone who had worked on Danone GMO yeasts at the Pasteur Institute.[7] He proposed his services to Monsanto—to its vice president David Stark—going so far as to ask him for damning information on me to take to the French government.[8] This last had just asked the commissions—who were themselves stoked with informers and lobbyists favoring the authorization of GMOs—for a second opinion of the Seralini study.

September 22. (MONGLY02080011). Eric Sachs toned down the insults in the draft of Henry Miller's *Forbes* piece, and also admitted that there was no error in the study's experimental protocol.

September 26. (MONGLY02063095). These gentlemen wanted to send a letter to Wallace Hayes, the compromised editor of *Food and Chemical Toxicology*, summarizing the reviews of the study they had collected. But they needed "authors" such as Bruce Chassy to form a large group so that the company's name wouldn't appear. He helped to find those whose labs had been financed by Monsanto all this time, particularly in the network AgBioChatter created just for support, and for this purpose to sharpen arguments for the public. They managed to recruit up to twenty-five scientists from fourteen countries. Some of the scientists had a bone to pick with Hayes for having failed to prevent the publication of the study. Their orders are clear: They must hide the fact that there are members of Monsanto in their group and make their reactions seem spontaneous. They put this in writing to each other.

In reality, their statements had to be pulled from them like teeth (or with dollar bills). They wanted to have them translated into French to have them published in *Le Monde* (MONGLY069909042). If Monsanto's arguments had held water, they would never have worried about putting their own name to them.

Under the Missouri sun, things are heating up at the Monsanto headquarters. It's 2018 and the scientist Charles Benbrook of the University of Washington, a former government pesticide regulator, has delivered a 244-page,[9] minute by minute study of the Monsanto Papers. It features the retired Bruce Chassy who was financed by Monsanto, Eric Sachs,

Heydens, Hayes, and Saltmiras. They were glued to their monitors across the country, getting more worked up by the minute, already seeking to "encourage the retraction."

September 27. (MONGLY02065511). The collaboration between Monsanto and the editor in chief Hayes was going full steam ahead, and Hayes asked Saltmiras to provide him with more reviews by different names. As he puts it: "The more the better." He was getting ahead of himself. Sachs admitted that there was no error, and yet, the subject line of the email is clear.

FROM: Sachs, Eric [AG/100]
SENT: Thursday, September 27, 2012 01:13 PM
To: HEYDENS, WILLIAM [AG/1000]; VICINI, JOHN
[AG/1000]; SALTMIRAS, DAVID A [AG/1000];
LEMKE, SHAWNA LIN [AG/1000]; GOLDSTEIN,
DANIEL A [AG/1000]; HAMMOND, BRUCE G
[AG/1000]; HELSCHER, THOMAS M [AG/1000]
Subject: RE: Help Take Action AGAINST Seralini
Study—Write to Journal Editor TODAY

September 28. Monsanto's excited minions wrote a collective letter. A portion of it appeared in the magazine *Marianne*. I attacked them for libel. The lot of them made up the so-called "international scientific community" in the name of whom they claimed to speak. It was a complete fabrication: As if there really was such a thing in a discipline that is by nature contradictory. The story was picked up by talk shows with remote-controlled hosts (like the ones from the magazine *Sciences et Vie [Science and Life]* and many others), who, in the name of neutrality, talked on without ever asking me my opinion.

It was early October. Following a meeting with Monsanto, Goldstein shared his notes. The meeting produced a to-do list that would lead to the retraction of the study. Action had to be taken quickly or the regulatory agencies could start banning the products.

Despite all this to-do in Saint Louis, Monsanto didn't get its piece printed in *FoodChemTox* until March 2013. This time it was written by the familiar names of Hammond, Goldstein, and Saltmiras writing as employees of Monsanto, claiming they had no conflict of interest in the whole thing. There was a double standard here: I supposedly had a conflict of interest because I was associated with environmental or public health groups and explained the results of my studies to them, while those who associated with multibillion-dollar corporations had none. We most definitely do not live in or share the same world.

It took another year, during which the journal did its best to figure out which arguments to use, but in November 2013 the retraction did in fact happen. I republished the study in 2014 with a European publisher. Meanwhile, I heard from several scientists and scientific reviews writing to me to express their disagreement or shock at the way Elsevier was behaving.

Most but Not All Are Rotten

Richard "Monsanto" Goodman was a researcher who created very toxic genetically modified bacteria—not a negligible fact for the rest of this story. In 2012 he suddenly found himself reporting to Hayes as the international editor for biotechnology at *Food and Chemical Toxicology* (they had fired the European editor who had originally published my article). He happened be the very person who had put together the regulatory files on transgenic maize for Monsanto without testing it over time. He was also the one who had Elsevier sign the press release retracting my study.

In the face of such an unprecedented scandal in the scientific community—let us not forget that according to Hayes himself there was no fraud or error in the study[10]—researchers the world over wrote to me in support and sympathy, suggesting I republish elsewhere and even organized a petition. Indeed, they knew all too well the laxity of the regulatory agencies when it came to evaluations of GMOs or chemical products. Monsanto's final goal was clearly to erase our research from the records of the official regulatory agencies in order that the IARC and the EFSA (European Food Safety Authority) would be unable to even cite it. Science editors from Springer, the large publishing group that rivaled Elsevier, suggested I republish with them. But as it turned out, the journal that published me, already in Monsanto's grip, anticipated this possibility and wanting to keep the rights, wrote to tell me that they had retained copyright to the article, making it impossible for me to republish it elsewhere. The goal of the retraction, and of absconding with the copyright, was to bury the piece forever. They wanted to beat it into oblivion along with its raw data and the entire corpus of related knowledge on the toxicity of Monsanto's products, GMOs, and Roundup.

Ready to blow a fuse, I answered that naturally I would not respect their decision and would deal with the legal consequences, as would they. However, my new prospective publishers feared going up against Elsevier in this manner, citing the enormous editorial group's power.

We were now in 2014. The head of Elsevier's legal department wanted to see me in person. He seemed of a better disposition than Hayes, who I learned was going to be changing jobs. I also learned that Goodman had returned to his previous stomping grounds, scot-free. The former editor who had originally accepted my article was being promoted to editor in chief. And to sugarcoat it all for me, they were going to offer me the job of editor in chief of another scientific journal. I answered them in short order that I couldn't work with a dishonest company. They went on to suggest a way for my article to be republished. All I would have to do was change the order of the figures (between the effects of the GMOs tolerant to Roundup and those of Roundup alone), and also reverse the title. As it currently read, the emphasis was on the tumors and toxicities provoked by Roundup on the rats—information Monsanto wanted to pass off as "natural."

After making it past another round of reviews, my study was finally republished in a new journal, in open source, *Environmental Sciences Europe*. The publisher was willing to help keep the debate alive. Doing this meant that the journal accepted to populate its website with loads of raw data—the minute details of the rats' blood or urinalyses became freely available for consultation by other researchers who could redo their own analyses or reproduce the experiment, and thus provide second opinions. You could try as you might to find the same data on Monsanto's research.

After that, from 2015 to 2017, we collaborated with King's College in London to confirm and publish our latest more detailed results, obtained with the most cutting-edge technology, with the prestigious group Nature (*Scientific Reports*). The lobbies came roaring back to harass this journal. All in all, despite some procrastination, other publications have come out confirming our results. The rest of the media still don't take our studies into account.

Chapter 8
THE DARK SIDE OF THE SHADOW ARMY

In which Monsanto's network kicked into action in the face of serious threats to its business (never mind to our health) and when the Bella Donna began to sing.

In the company's 2013 annual report, it was noted that one of its employees, David Saltmiras, prided himself on having successfully enabled shady practices and on having served as "leverage on the publisher" (MONGLY01045298) against my research. He served a powerful role as the arbiter of which articles about Monsanto's products got published, while keeping the company's name out of the fray. What was said about Roundup the world over was thus under Monsanto's control. Behind that control was a certain Donna Farmer—as we gradually found out.

The Bella Donna
Let us draw our attention for a moment to this key player, who, as we will see, will be charged with personal liability further down the road. Those who directed her or endorsed her wholesale massacre of science—in the image of the politicians who knowingly or unknowingly were duped—should meet the same fate.

Donna Farmer played an essential part in the business of confounding Roundup and glyphosate since at least the year 2000. She was Monsanto's lead toxicologist, in charge of regulatory affairs. She should be found guilty of hiding the dangers of Roundup. As it turns out, during her testimony[1] she admitted to never having tested it over time. This is something she should have done; it was logical.

When she joined Monsanto in 1991, she was really joining a factory of lies. She would have to become impregnated with the company culture. As early as 1984, Monsanto's scientists showed that carcinogenic byproducts attached themselves to glyphosate (or added themselves to it) in the formulants. The "N-nitroso-" compounds are among the most powerful carcinogens we know (MONGLY00925905). Despite this fact, thirty years later, in 2014, they are still present in the product (MONGLY03549275-80).

Monsanto's experts are careful not to draw attention to the toxicity and carcinogenic nature of the chemicals present in Roundup. From nitroso- to 1.4-dioxane (MONGLY01041300), and including formaldehyde (MONGLY0090361), they know these "surfactants" well. Their name suggests that they remain on the surface of organisms, when in fact they are very penetrating. In regulatory parlance, the terminology is constantly falsified; meanwhile these known carcinogens make it through the gate (MONGLY009200065). Among the formulants are some oxidized—in other words, burned—petroleum derivatives that Monsanto's team tried to have classified as inert and confidential, all to avoid having them listed as active ingredients.

They were really serving it up. They may as well have been saying, "try a little alkyl-amido dimethyl propylamine—here's to your health!" A new word was introduced to classify yet another poison hidden in the pesticide's formulant: "humectant" (MONGLY01832749). In this case it refers to the very well-known and omnipresent ethylene glycol. Daniel Goldstein knew it was toxic for children as early as 1999. One more "inert" substance to sweep under the rug.

What's more, as we emphasized earlier, Monsanto's experts went so far as to feign ignorance of our studies of arsenic, published back in 2018,[2] even as they simultaneously fought to discredit them. When we discussed the topic in a previous book:[3] *The Great Health Scam*, we thought that this veritable octopus of the corporate world might be ignorant. There is now no doubt, thanks to the Monsanto Papers, that this was not the case, that the creature is an expert in dissimulation, and that it is more intelligent and organized than it appears to be, all the better to hunt and expand its territory. To persevere and prevent its operation from derailing, it has a strategy: change the subject.

Even after the company was fined $2 billion for the cancers its products caused, the media and the NGOs were abuzz with the same old debate Monsanto created over glyphosate alone. Only the lawyers, especially the American ones, stayed focused on the real matter at hand, came in for the kill, and walked away victorious. An independent review[4] showed that the authorized products in the pesticide mixtures were linked families of molecules such as ethyl oxide and propylene oxide, not tested over a long time period. A rapid internet search shows that ethylene oxide became of consequence during the First World War, when it was used to make ethylene glycol (as a refrigerant) and a chemical weapon (mustard gas, or yperite). The genotoxicity of this product was documented as early as 1968. Its cousin propylene oxide is a well-known carcinogenic mutagenic reprotoxic chemical; propylene is the second most prevalent molecule of petrochemistry, with a worldwide annual output of around 70 megatons.

In France, in 2019, the Tumerelle law firm was alone in successfully denouncing the hidden poisons of Roundup in the Court of Justice of the European Union.[5] The Court put it to the member states to force all manufacturers of pesticides to declare all of their active ingredients. Not just glyphosate, but also arsenic for example, which was a pesticide, or tallow amines, derived from petroleum and pig fats. At this point it is clear, these products are far from being classifiable as "inert," as the company got away with calling them, and they wouldn't be able to legally remain "secret" for much longer, especially if the press and the public, or even the NGOs who focus on pesticides, took an interest in them. But instead of this happening, the NGOs and the government stayed focused on the least dangerous component of Roundup besides water—glyphosate. Three cheers for the Bella Donna (Farmer) for her excellent work!

Donna Farmer was the principal architect of the glyphosate mess. If any confusion still exists despite the outcomes of the court cases, it is thanks to her maneuvers. I invite you to observe her face during the hearing.[6] You can also ask her yourself: You'll find her in the Monsanto offices of her new professional address at Bayer, where she still holds the same job. In the March 2019[7] issue of *Regulatory Toxicology and Pharmacology*, she stated her position on "polyethoxylated tallow amines." Speaking for Bayer, she seemed to have changed her mind about them. She concluded, according to the synopsis, by saying: "These surfactants have no significant effect on human health." A little greenwashing for the new boss.

She would become famous for this line, pronounced in the California court during her testimony: "We cannot say that Roundup does not cause cancer. We have not performed the necessary tests on the formulated products." This was already true in 2003, and Donna admits it (MONGLY00922458) and she admitted it once again in 2017 (in her legal deposition). Our team had done those tests on human and animal cells. That's why they couldn't stand us and we couldn't stand them. "Seralini is a threat to POEA,"[8] one of the families of compounds added to glyphosate. Of course, they are the ones who are the real threat to public health!

Brent Wisner said this about Donna Farmer: "She is chief toxicologist for Monsanto, someone who knows Roundup better than anyone else in the world can claim to." Donna Farmer's public confession shocked the jury and was extremely notable. It was much more crucial than it appeared: it underscored the company's communication strategy that once again, tried to confound glyphosate and Roundup—let us not forget that glyphosate alone was tested at the regulatory level by Monsanto over the long term. And yet, not one person has ever used glyphosate alone in the fields or in the garden.

Professor Robert Bellé confirmed this: "What creates the confusion is the not trivial choice of words." Together with the Paris-Sorbonne University, this researcher from Brittany had demonstrated the potentially cancerous nature of Roundup, and not of glyphosate itself, in an experiment conducted in 2002[9] on sea urchin eggs. This was why he was blacklisted in Monsanto-Bayer's secret file (as am I).[10] He filed a complaint with his university. He went on to say: "As long as the European Food Safety Authority refers to glyphosate, it will continue to say it is harmless, which is not completely untrue [. . .] the problem is that it is Roundup that is spread over the fields."[11] In fact, the EFSA duplicitously and unabashedly cut and pasted Monsanto's ghostwritten reports on the harmless nature of glyphosate as if they had written them themselves.

Ms. Farmer is one of many phantom authors of "scientific" literature who wrote about the so-called inoffensive nature of glyphosate. She was published in scientific journals controlled by the industry where she already had had a hand in cleaning up any inconvenient mess. But her name was removed at the last minute by her boss: The harmless nature of glyphosate would appear more genuine if the articles were not written by someone from Monsanto (MONGLY00919381).

Donna Farmer did her homework and discussed the mutagenicity or toxicity of the formulants internally, the ones we would call hidden poisons, like tallow amines for example—and deviously called them "surfactants" or even "adjuvants." This was all corporate-speak that little by little became regulatory parlance to make it sound as if these products were just there to help the glyphosate along a bit, when in fact they were way more toxic; all the experiments prove it, even the one conducted by Monsanto.

The Losing Three-Card Monte of Hidden Poisons

Would you prefer, dear reader, a calming bedtime story to help you fall asleep, the kind Monsanto gave the regulatory authorities, or one that portrays the fake science? Here it is: One fine morning between 2002 and 2004 (MONGLY06424476 and MONGLY06409924) Dr. Charles Healy (Chuck) had the brilliant idea of conducting research to prove that the surfactants of glyphosate didn't enter the digestive system. This would definitively have shielded the company from regulatory oversight, and the surfactants could be used as examples of other "inerts." No such luck. It turns out that the rate of absorption our wunderkind discovered was enormous: at least 56 percent, which is of the order found with antibiotics.[12] His conclusion was "scientific": "None of this should be reported." Then: "It's not what we were hoping for." One wonders what they were hoping for. I pondered this as I fell asleep. They would have to find other ways. When the rats in their experiments died (MONGLY0672256), they never attributed it to the treatments.

Donna Farmer finally succeeded in having these toxics classified as "inerts," and therefore, they would remain undeclared (MONGLY06414231). Monsanto went as far as to appoint a person in charge of "protecting the tallow amine-based formulations in Europe." (MONGLY06414231). In France however, this tactic failed. The minister Ségolène Royal wanted to ban them, which she did in 2016 based on the results of our studies, the very ones that prompted Monsanto to declare that "Seralini is a threat to the POEA."

Monsanto's experts still hoped to sell their products into Germany—and elsewhere—and declared the following (MONGLY03401522): "We stand by our position of non-toxicity.

[. . .] The French agency should not base its conclusions on imminent threats to human health, but on other considerations such as the principle of precaution. It is paramount that not all the efforts to ban the product refer to the risk to human health, otherwise this will extend beyond France and have an impact on the brand at the global level." Such a degree of cynicism would have even made Diogenes roll over in his barrel and says a lot about Monsanto's communication strategy even as they sat on their own internal studies proving toxicity.

These MONGLY were carefully selected among a myriad of others that confirmed these facts. We could stop there, but many more documents are available to the public on the websites of Baum Hedlund, and US Right to Know.

In January 2020, was the US government trying to clear its complicity? Was it trembling in the face of the army of lawyers representing 100,000 victims who would question its responsibility? Fifteen years after our first article proved it, the National Toxicology Program (NTP), a division of the National Institute of Environmental Health Sciences (NIEHS). woke up from a criminal sleep. The NTP published a (preliminary!) report listing evidence that the formulants in Roundup were more genotoxic than glyphosate alone. The only problem was that Monsanto had classified these formulants as confidential. Bayer was on the case in no time: "The majority of the ingredients are harmless." My teams had of course long since identified the exact nature of these toxic products, such as heavy metals like arsenic. In 2018 the French secretary of state even had to resign because he knew of this without being able to do anything about it. Donald Trump would defend the industry. But it didn't stop there.

Chapter 9
"A PAIN IN THE ASS"

_In which we discover that I'm not the only object of Monsanto's
kindly attention, or their pain in the ass. Reporters, international agencies, scientific experts, government officials—we're all
in the same boat. Meanwhile the company continues to coddle
their suppliers of hidden poisons._

"A pain in the ass." This is how Carey Gillam, who heads up US
Right to Know, is described in the Monsanto Papers. She is a
reputable journalist who used to work at Reuters. With her colleague
Gary Ruskin, she played a key role in obtaining, sharing, and
analyzing Monsanto's secret documents. Leemon McHenry from
Baum Hedlund did the same. Now the company was pulling out all
the stops to destroy her reputation.[1] Internal communications
showed that she got the employees all riled up, and that they spoke
about her with profanities. The tough and determined investigator
defended herself by publishing more and more stories, like the one
about the Welsh scientist James Parry.

Just as in the game of three-card monte, the industry didn't
always win, and at this point companies were frantically talking
to one another, trying everything they could and throwing tens
of millions of dollars at the problem.[2] We were not the only ones
celebrating. You may recall James Parry of the University of
Wales quoted at the beginning of this book. He is one of the
world's top toxicologists. Monsanto managed to get him under

contract, but in 1999 the researcher discovered the mutagenic effects of Roundup, and the company buried his report (MONGLY01314233-270). Any findings under this type of contract were legally the property of the funders. Should the recipient disclose any confidential information without the express permission of their financier, they could say goodbye to their funding and fear legal repercussions.

As they struck out with Parry, who discovered as we did the toxicity of the product, the company sought to align itself with someone with fewer "final intentions" (MONGLY00878595-97), despite the fact that Parry was considered an objective top researcher. William Heydens, the other major "psycho-toxicologist" for Monsanto (aka Bill, whose actual degree was in psychology), decided in the end that it wasn't worth redoing or prolonging these revealing studies (Monsanto picked and chose the results that suited them, the ones about non-mutagenic glyphosate, while conveniently leaving Roundup out of it on the grounds that it would require too much time and money [sic] MONGLY03734971), nor to do any further studies to challenge Parry's position. This very same Heydens, however, wrote in 2015 that "the surfactants within the formulations [. . .] play a role in the promotion of tumors" (MONGLY01183933). He knew. So it was the return of the repressed. The old problem of the formulants had resurfaced. A new deployment of energy and money was needed to bury it once again.

In 2010, as Parry's studies were falling by the wayside, we demonstrated the very same thing: glyphosate-based herbicides with formulants are mutagenic. Our study, however, was done in 2009, and on human cells. Our first attempt at publication failed, much to the delight of the company of shadows.[3] Even though it did not get published the first time, they kept tabs on it as if they had been stalking it for quite a while. We did finally get it published in a scientific journal that was a bit more under the radar.

This is when Donna Farmer readied her bevy of critics and had their reviews of the piece signed by Williams (MONGLY00919381).

The IARC Circus

They will really stop at nothing. When the International Agency for Research on Cancer (IARC—a division of the WHO), specializing in cancer risks, took an interest in Monsanto's lead product, the company went on the war path, fighting the scientists and the agency itself by any means possible. In light of the controversy, seventeen experts from eleven countries got together and met from the 3rd to the 10th of March 2015 to analyze all the public data on five pesticides, including glyphosate. My work on human cells was taken into account and quoted but not so my work on rats, as it didn't focus specifically on cancer. There are other ways that pesticides can kill you, but those were not the purview of this particular agency.

The researchers at the IARC took the courageous and innovative stance of taking into account the public and published research on the topic, while ignoring the proprietary and secret data that industry served up to the agencies for the purpose of getting commercial authorizations. They took an interest in the actual facts from the field, including epidemiological research on patients exposed to Roundup as it is marketed; in other words, with the formulants. It is important to underscore that this was not the mandate of agencies like the EFSA, who were expected to rule on glyphosate alone. At least that is what they said. The European Court of Justice would interpret the law differently in 2019 due to our work. This erroneous approach is the source of all the confusion. It only took glyphosate's toxicity into account. This approach goes as far back as the post-War period and was put into place by lobbies for the industry.

In 2015, the IARC classified glyphosate as a "probable carcinogen" (class 2A). They also concluded that there was "strong evidence" of genotoxicity both for "pure" glyphosate and "for the formulations based on glyphosate."[4] Henry Miller, the so-called "author" with the red sedan, was again first in line to destroy these findings in an article in *Forbes*. For this he touted his position as an "independent scientist." As usual he asked his influential Monsanto friend Eric Sachs for a "substantial draft" of the arguments (MONGLY02063611, MONGLY02063572). Even the Bella Donna took part—the usual suspects!

Donna Farmer asked her colleagues how to go about paying an expert from the IARC $12,000.00 (MONGLY02816607) so that he would fail to find anything to do with genotoxicity. Monsanto would do the same with John Acquavella, a professor at the University of Aarhus in Denmark. He was paid $20,700 the same year, on top of his salary (MONGLY-03934897), and even more over the years, as his CV includes lax reports done for the IARC as early as 1993. He didn't hide that he was under contract with Monsanto.

One can only imagine the number of Acquavellas the company could have paid with the $17 million they devoted to discrediting the IARC for the year 2015 alone. Monsanto's director of public relations revealed this during the hearing.[5] Could this dedicated sum also have paid off the Reuters reporter, Kate Kelland, in order that she diffuse Monsanto's case against the IARC in Europe? She does seem to work with the company on a regular basis. This particular time, she was under strict orders not to reveal that the piece originated from Monsanto. The goal was to undercut the president of the IARC under the pretense that he had hidden some facts that would have cleared glyphosate's reputation.[6]

During Monsanto's weekly crisis control meetings on glyphosate, the company had anticipated that if the formulants were classified as carcinogens, a flurry of patients' lawsuits would follow as early as May 2015 (MONGLY03315608).

The goal of this book is not to study the whole IARC affair, but to shed light on the methods used by Monsanto as soon as its profits were at risk, while they failed to ask themselves if their products were safe. Meanwhile they were also covertly attacking politicians, namely Ségolène Royal and the Secretary of State Nicolas Hulot in France. Hulot wanted to take action against the products in the formulations of Roundup, and after he resigned from his post, accused Monsanto of trying to damage his reputation via a Belgian bureau (which was actually a PR firm in disguise).[7]

Elegance According to Monsanto[8]

The investigative reporter Carey Gillam and I appear in the company's internal documents in turn as "pain in the ass" and "persistent bugger." Evidently Monsanto projected their chronic annoyance with their opponents in rather vulgar terms. The other mind-boggling thing was the anti-scientific behaviors that clashed with the ethics upheld by the scientific journals. Since 2005, I have been careful to explain to independent publishers that Monsanto, or the regulatory agencies (since I have demonstrated their faults), must need be excluded from any evaluations of my publications, as they have serious conflicts of interest—just as one can denounce a judge or a juror in a case where they have preexisting ties to the case. But Monsanto has long infiltrated the scientific publishers and spied on my activities as a researcher to try to keep me from publishing my results. They will stop at nothing to do this, even allying themselves at times with the competition, such as was the case with Syngenta. They expended considerable efforts to this end, as demonstrated in 2011 (MONGLY03081997).

In January 2011, following the news of our disturbing studies, William Heydens (our psychiatrist, Bill), a cog in the wheel at Monsanto who had worked for the company for thirty-three years, spoke up. In an internal memo addressed to Daniel Goldstein, another Monsanto colleague (if you'll recall, he is a

pediatrician—presumably looking after the welfare of children—
and a clinical toxicologist) Heydens referred to me as a "persistent
bugger" due to the fact that we continued to publish about the
hidden poisons in Roundup, mixed with glyphosate.

January 2014: The spying continued, and Monsanto got a heads
up from Vicini and Sachs that I was about to publish an article
about the toxicity of the formulants of pesticides
(MONGLY07035266). Also in 2014, the lobby in defense of the
biotech industry, Croplife, was hired to propose topics to Monsanto
detrimental to my reputation. But they never signed their name to
them, and desperately sought experts from the regulatory agencies
to do so. In 2016, a researcher by the name of Karl Haro von Mogel,
from Biofortified (an educational association as they described
themselves, that was so-called independent) and AgBioChatter,
was paid to attend a conference I was speaking at in San Diego,
CA. I was to receive the Théo Colborn Prize from the Environmental
Health Symposium. There were four hundred doctors or other
medical professionals in the room. Passing for a busy toxicologist,
Karl was a youngster, with only one publication to his name on the
metabolism of corn starch. He was probably looking for a job. He
followed me around, never introducing himself, and signed up for
all the panels I was on in order to report back on what I said to
Monsanto. This was relayed by a researcher connected to the company, Wayne Parrot (MONGLY07005764). Karl demanded more
funds in order to write articles against me. He wrote one three-page
report on the presentation, explaining my claim that the toxicity of
the formulants of Roundup (100,000 times higher than that of glyphosate), baffled his brain, but that he felt that it would be able to
withstand the attack.

The Secret Ingredients
The most shocking of all our discoveries was that Monsanto was
well aware of its crimes. Let's go back for a moment to 2010 and the

damning email conversations Heydens had that were admitted as evidence against him by the California court during the 2018 trial. They show why he was so preoccupied with polyethoxylated tallow amine (POEA), tallow amine, and other rotten chemicals that we know are byproducts of petroleum and animal fats or of the remains of recycled skeletons. It was confidential (MONGLY02062439): Monsanto's experts and scientists rushed to the politicians and regulatory officials in order to prevent a domino effect, and to make sure that the market for fertilizers and other corrosive and mutagenic detergents based on pesticides did not collapse. They focused particularly on Brazil, North and South America, Europe, Scandinavia, and Germany. They didn't want the carcinogenic and toxic powers of these chemicals revealed.

One wonders who would make these chemicals? Still the emails (again MONGLY02062439) revealed not concern about the health risks of their products, but concern for the future of the Zanussi company, a chemical company that makes detergents and home appliances. They were also concerned with the plans being cooked up by another chemical company, Akzo, who may have been willing to put up a smoke screen as to the pseudo-innocuity of the chemistry. The question was how quickly they would be able to replace (or not) these poisons by others whose regulatory toxicity would stay better concealed, and only in products that were glyphosate-based. They decided to get to work on this, so that no one would bother them over their presence "in other pesticides or personal health care products."

The list of companies that regularly resold products to each other was long: There was Akzo, or their subsidiary Nouryon, and its "brands"; Morwet who made "dispersants"; Adsee, who made adjuvants; Agrilan, polymers; Armid, solvents; Clariant, Hoechst, Aventis. They all synthesized or isolated products based on petrochemical residues. They had been obtaining the chemicals they needed since before World War II from companies such as

Rhône-Poulenc. This was where the true basis of the petrochemistry of pesticides entered the mafioso's game.

When there was a fire at Lubrizol, the Rouen petrochemical company, they found these same chemicals in the atmosphere and in the food chain. The state didn't consider that this presented any chronic danger as the doses of dioxin and other known carcinogens and hormonal disrupters were so-called safe. Lubrizol's products were destined to be additives to gasoline or food. They've sidestepped questions about these dangerous liquids since 2002, the year the EPA asked them to come clean.

Heydens wrote the following to Donna Farmer "[. . .] we are in pretty good shape with glyphosate but vulnerable with surfactants" (as to their toxicity). "Glyphosate is OK but the formulated product (and thus the surfactant) does the damage." (MONGLY00885526). Instead of testing and replacing the chemical products (but when one intends to create a product that kills, how could it not kill? Can a poison cease being a poison?), they spent their time avoiding the authorities, or bribing them, even as they were shooting the messengers who discovered their flaws.

This happened to us several times, but their practice of ghost-writing and the other shenanigans to make people believe our research was flawed came to a head, as we saw, in 2012. The Bella Donna, Ms. Farmer, had a smaller part to play in all this. But now, who should enter the stage but the US government.

Chapter 10

THE CORNERSTONE OF THE BUSINESS OF GMOs

In which nations competed among themselves to see who could have the most influence, highlighting the impact our 2012 study had around the world, while Monsanto organized a game of golf with the US government. And how governments and agencies found themselves forced to cover up their laxism by having phony studies performed.

In 1980, the Supreme Court of the United States became the first in the world to rule in favor of the patenting of a living genetically modified organism (GMO), setting off a legal revolution that would affect the food supply, the environment, and the pharmaceutical industry. Once it had been genetically modified, any living thing containing a commercially registered patent could henceforth become private property. The biotech company Genentech, which sold patents on genes, was able, for example, to increase the value of its stock fifteen-fold in one afternoon, congratulating itself with this remark: "The Supreme Court has guaranteed the future of our country's technology sector."

Subsequent administrations continued to support and promote GMOs, as did the Bush family. Dan Jenkins, in charge of relations with the US government for Monsanto, played golf with the Bushes. When I published the 2012 study that proved the toxicity of a GMO modified to absorb (and thus contain) Roundup, I inadvertently started a war between nations. Russia banned the GMO as early as the next day. Meanwhile Monsanto realized that the only way to stop the war between nations was to declare war against me first.

As soon as it was published, Monsanto and Dan Jenkins were on thin ice (MONGLY00978886). Saltmiras went on record as saying "Seralini set off a firestorm and can now declare a modest victory." We found out that two days earlier there had been a teleconference between countries of Asia, the Pacific, the Americas, Europe, and Africa, to decide the next steps to take. In response, the Monsanto team came up with a strategy to retract the research and to prevent any further such studies. The next day, Goldstein reiterated the key measures to be taken to sixty-two of Monsanto's CEOs around the world. First, Dan Jenkins and the head of operations in Asia, Harvey Glick, emphasized the importance of the retraction that would "take away my liberty to operate." They also quickly moved to put pressure on the journal that published me, *Food and Chemical Toxicology*, to retract my study. They saw no other way to fix the problem.

Richard Goodman, the renowned specialist of transgenic bacteria who worked for Monsanto between 2004 and 2005 and is now a professor at the University of Nebraska, was copied on the memo, as was the editor William Hayes. Following the storm my article provoked, they replaced the European editor of *Food and Chemical Toxicology*, José Domingo, with Goodman. Hayes had fired Domingo for accepting to publish my research. The new editorial team asked me to resend them the raw data from my study, as if there hadn't been any previous review. It was inadmissible that this former Monsanto employee, who labored on behalf of the very same genetically modified corn I had written about, should be allowed to rule on the subject.

I complained to Elsevier, the journal's parent company, about the conflict of interest of having Goodman review my data. The legal department tried to reassure me. But the whole thing was starting to smell fishy, and in order to protect myself and to be treated fairly, I requested that there be likewise a second opinion of the raw data Monsanto had used in order to obtain authorization for the same GMO and for Roundup. My supporters even organized a press conference at the European Parliament in Brussels during which they submitted a USB key containing all of our raw data in order to push Monsanto to do the same. We're still waiting for it. One can only conclude that the authorities were lax and careless.

The study led to questions from countries around the world, but Monsanto's experts wanted to avoid at all costs having to perform long-term studies on the pesticide as it was formulated and on the GMO, even though they knew that interactions within the formulations and with the GMO itself were possible. The company's toxicologists stopped there rather than experiment on the root cause of the problem, and will one day have to justify this strategy. The European Commission, in the face of the controversy, finally budgeted 8 million Euros of public funds to have a study done over two years with rats fed the GMO, supposedly to double check my own study. Good.

But there was something not right about it all. The European Food Safety Authority arranged for Roundup to be excluded from their study, and named Pablo Steinberg, a member of the International Life Sciences Institute and of the German Federal Institute for Risk Assessment (BfR),[2] as well as a consultant for Danone, to lead the project. In his article,[3] Steinberg states that he observed disease in all the rats, of the very same order as the ones that were in the control group. The control group, which was there to show the difference between a healthy diet and the product being tested, was in fact given food contaminated by a dose of glyphosate similar to that in the Roundup used on crops, or 300 to 1,400 times higher than the dose I had studied. In addition to other pollutants, their food contained arsenic. This was plainly a ploy to make me look bad. I would later denounce it in a detailed scientific article.[4]

But back to my original editor, Domingo. . . . As his judgement was called into question, the journal got a second opinion. The study was judged flawless by the academic leadership of the journal. But this didn't stop the retraction, which took a year. In the interim, Monsanto's thugs had begun to lose confidence (MONGLY02719740). On November 30, 2012 at 6:44 a.m., the boorish Heydens piped up: "Even if it takes time, Seralini will be back" (in other words, will continue to publish, to their great displeasure). And Sachs wondered: "Will there be a light at the end of the tunnel?" They thought they had finally found it in Steinberg's shady study.[5]

Chapter 11

FATE OR A BULGARIAN UMBRELLA?

In which a black tie and evening gown show up at London Hospital at the bedside of a suffering Seralini, while back in Saint Louis, in Crève Coeur, makers of genetically modified bacteria hold a juvenile celebration.

In the summer of 2013, Monsanto told the press that the company was throwing in the towel and would not be introducing any new GMOs in Europe. But the context in which they made this announcement was anything but normal: The authorization of a new transgenic corn, the TC 1507, was slated for discussion the next fall. Arguments were erupting in parliaments.

I had been invited to present my controversial results in Great Britain at the very beginning of September, in order that the Scottish, English, and Welsh parliaments could form an opinion. I was scheduled to spend two days in Edinburgh, two in London, and two in Cardiff, each time in a university, in parliament, and in front of the press. My article was not yet retracted—this would only happen in November—and the discussions around it were heated. I felt very welcome in Edinburgh, as the Scottish Parliament was already against GMOs, and the English position.

I was met at the London airport by the organizers and informed them I wished to take a taxi. One of them was insistent that the train and the underground would be much faster if we wanted to make it to the welcome dinner on time, and said that besides, we were too many to fit in a cab. I reiterated my wish to take a cab, to have a moment of peace and quiet. They insisted on the alternative, especially one of the organizers in particular, whom I didn't know. I trusted some of the others. The underground was packed, and the heat was unbearable. A few stops before my hotel, a suitcase banged me just under my left knee and I felt an unpleasant sting. I didn't pay any attention to it.

The dinner went well, despite the fact that the cooking was British. In the shower I noticed redness where the suitcase had banged into me. I still paid no attention to it. The next morning, I woke up with the symptoms of a bad flu: fever, fatigue, and aches. I was preparing for a late-morning meeting with some members of parliament and thought to myself that an antipyretic and some antibiotics would be welcome, given the political and environmental stakes: to reestablish the truth about Monsanto against a backdrop of British support for the new GMOs.

The parliament listened carefully to my presentation and asked many questions. I left the room sweating from the heat and fever and agreed to go to the parliament infirmary. They took my pulse, my blood pressure, and asked me what I was feeling: pain in the leg and lower back, and fever. The nurse lifted my left pant leg and as she pulled it back down, she said I must go straight to the hospital, on the other side of the Thames, for antibiotics. I grumbled but I was already being whisked away in a wheelchair and the parliamentary assistant accompanying me reassured me that it was for the best. The wheelchair fit in the back of the big London taxicab, which, to my surprise, had no back seats. The journey ended at the emergency room where I hardly had to wait. The political aide knew how to open doors.

Somehow, I ended up in a green hospital gown in a space surrounded by dark curtains with a woman in a tight black dress and an elderly grey-haired gentleman in a bow tie. *So British* (notes my coauthor). I thought I must be dreaming. A number of people in scrubs stood around them. They handed me a waiver to sign, that would allow them to operate immediately. I wanted to leave, didn't have any idea why I was there, and was feeling tired and confused. I felt my nudity under the gown.

The lady asked me if I knew why I was there. It all seemed blown out of proportion, off-kilter, constraining, useless. I explained that I needed to give a talk at the University of London that very evening and that I was supposed to meet an authority on hormonal disturbances caused by pesticides. Also, that I had a flight to Cardiff the next morning. The woman playing the role of doctor in the tight black evening gown comes up to me: "Forget all of that. They called me away from a performance I was attending with the chief surgeon (pointing to the bow tie) because you are in danger of dying tonight. You must have surgery now. Have you seen your leg?" My head was spinning, I wanted to pinch myself. What on earth were they talking about. I took a furtive glance at my leg. My left leg was black and yellow with pus from the ankle to the knee. I was scared. "You won't amputate it?"—"No it's too late, the infection has spread to your kidneys and lungs. We're going to take out as much of the gangrene as possible and we'll do a transplant to rebuild your leg." All of this happened in twenty-four hours!

I was not reassured. I signed the waiver, in shock. The last thing I saw was a nurse hooking me up to a drip, and then I lost consciousness.

I woke up with tubes coming out of everywhere, a mechanical lung, a feeding tube, and even a tube in my penis it seemed. I could not speak, nor could I lean over to look at my leg. I learned afterwards I was in the intensive care unit. From the end of the hallway, I could see them wheeling out the dead, their faces covered. Two male nurses took turns at my bedside day and night for three weeks.

They observed my heartbeat on the screen and showed me my blood oxygenation level, constantly reminding me that it needed to be higher. I communicated by writing on a slate. I was constantly dying of thirst. I wrote "WATER!" They told me that I was getting hydration intravenously and that they couldn't give me any by mouth because I could choke. Following the surgery the fever returned, reaching new heights. They called my loved ones to my bedside. They told me they had thought it was the end. I was in a comatose fog most of the time. I wanted to speak to my children and to rip out the intubation. They finally took it out so that I could speak. I gasped for air and took a deep breath that oxygenated my blood.

Weeks later I learned from the discharge papers that I'd had an antibiotic-resistant streptococcus, previously unknown, that had ravaged me in less than twenty-four hours. No one else in London was known to have been infected. They couldn't trace its origin after weeks of searching. A deep scar on the upper left side of my tibia marks the place it entered my body. There is a rough groove fifteen centimeters long where they grafted skin from my thigh. After I left the intensive care unit I was transferred to Guy's and Saint Thomas' Hospital to recover. Five months after I was discharged, I was finally able to walk normally again and to regain proper kidney function.

My job is to rummage through all sorts of scientific research archives and papers. A long time after my trauma, I tried to understand the virulence of the antibiotic-resistant bacteria that had infected me so quickly. I went online and found that one of the best-known sites belonged to the University of California San Francisco (UCSF). In the "Industry Documents Library" I entered the key words "antibiotic resistant human streptococcus." I got 731 results. At the top of the list there was a review by specialists, entitled "Human Safety and Genetically Modified Plants,"[1] that presented the technology used to transform bacteria by inoculating them with antibiotic resistant genes. I was taken aback and could not believe my eyes. The article was issued by the Monsanto

Company and was signed by D. A. Goldstein, with contributions from a certain R. E. Goodman, the very same person who was promoted to international editor in chief for biotechnology at *Food and Chemical Toxicology*. The link to my condition may have been tenuous, but it was such an unusual coincidence that it begged examination.

One year from the publication of my study, and still recovering, I thought the worst was over. I had been reassuring my collaborators that if *Food and Chemical Toxicology* had found the slightest problem with our raw data, they would have already exploited it.

In the Meantime, Elsewhere, It's Party Time

November 2013. Crash bang. Hayes announced that he was going to revoke the study: "Unequivocally, the Editor-in-Chief found no evidence of fraud or intentional misrepresentation of the data. [. . .] Ultimately, the results presented (while not incorrect) are inconclusive [. . .]. The retraction is based only on the inconclusiveness of this one paper."[2] This was after I had already answered all the queries received in a long article that the same journal had been forced to publish[3] and that it was not retracting!

According to Goodman, there were no errors, so the rats did indeed die from the treatment—the photos say it all—it's just that our conclusions bothered them. I would have to fight back against these mafia-style acts.

If I'm revisiting this whole episode, it's because on September 20, 2013, two months and a week before the retraction, the folks at Monsanto were already congratulating themselves and even created a prize for the staff entitled "I smell a rat—Response to Seralini." ("Scientists smell a rat," refers to the original title of Henry Miller's article in *Forbes*).[4] The nominees for the prize included Daniel Goldstein (the one who had developed transgenic bacteria that were multi-resistant to antibiotics) as well as Bruce Hammond, David Saltmiras, John Swarthout, and Eric Sachs.[5]

There were in fact ten categories of "awards against Seralini" that the employees of Monsanto were eligible for (how many for heaven's sakes were involved in these charades?). John Vicini asked Daniel Goldstein to confirm the award categories: Unsung Hero; Against All Odds; Breakthrough Innovation; Building the Team; Paving the Road; Moving the Needle; Newcomer of the Year— North America; Newcomer of the Year– International; Getting to Market (compromised experts); Partnering for Success. But what kind of success were we talking about, this 20th of September? The article would not be retracted for two more months. They still had doubts that the retraction would happen: On November 28, 2013, they wondered if it would actually take place (MONGLY01004347). In the end, the journal didn't publish the retraction notice until 2014.

In the meantime, Monsanto was thanking their teams for the shenanigans they were carrying out against science and against my person. Let's take a minute to let them describe in their own words just what they were doing, in retrospect, to share their celebrations and the talk that went with them. All of this would be taken into account in court in the 2019 California trial (MONGLY02546318, 2013).[6]

The Seralini study was a multimedia event that was designed for maximum negative publicity. The Monsanto toxicology team was mobilized to provide a rapid assessment of the technical aspects while the scientific affairs team helped organize third-party scientists briefed on the matter to react to the study. In all, it took six months for them to respond. The final piece included Monsanto's technical evaluation, a letter to the editor (longer than the original manuscript), responses by the Glyphosate Task Force[7] (allied companies), PowerPoint presentations, responses to numerous regulator inquiries, blog posts, and popular articles in the press. This was the result of coordinated efforts and synergies by people from multiple regulatory teams.

Why not, while they were at it, celebrate the poisoning of the world?

In 2019, during the Edwin Hardeman lawsuit (that Monsanto lost, forcing them to pay an initial sum of $75 million), this is what was said in regard to the article: Evidence demonstrates "that the Seralini story is central to Monsanto's failure to test its products as well as its efforts to manipulate public opinion."[8] At the same time, European journalists continued to argue that glyphosate was the lighting rod of the ecological debate, and they questioned heads of state over the delays in its eventual ban.

Certain regulatory agencies reasserted glyphosate's innocuity, going up against the conclusions of the International Agency for Research on Cancer (IARC). In addition, three American lawyers from the law firm of Andrus Wagstaff of Lakewood, Colorado (the litigation had spread across the United States), confirmed that "the testimony[9] reveals that Monsanto responded to the study by attempting to undermine and discredit Dr. Seralini, which is further evidence that Monsanto does not particularly care whether its product is in fact giving people cancer, but focuses instead on manipulating public opinion and undermining anyone who raises genuine and legitimate concerns about the issue."

A group of scientists forming an opposing scientific community condemned the retraction of the study.[10] But even then, the press only put credence in the "scientific community" formed by the agencies and the academies, who touted themselves as THE scientific community. PR firms and the lobbyists disguised as experts ceaselessly repeated themselves. The trust in this alternative scientific community, who unilaterally agreed that I produced fraudulent studies, proved so deleterious to my career and that of my collaborators, that we filed lawsuits, together as well as individually. I filed seven lawsuits between 2011 and 2017 as a civil servant (researcher) defamed during the exercise of his functions. I won seven times, and gradually learned, thanks to the Monsanto Papers, that it all came down to these mafiosos.

Chapter 12

SERALINI AGAINST THE "SCIENTIFIC COMMUNITY"

In which the president of a commission, France's pitiful "chief expert," gets sideswiped and reports back to Monsanto about his defeat in court, and in which we strip bare the company's international networks.

I was a member of the *Commission du génie biomoleculaire* (CGB), or biological engineering commission, from 1997 to 2008, under the direction of the geneticist Marc Fellous. As a student, I had heard people talk about Fellous as an established researcher. On the commission, he and I often disagreed over what I considered the lax criterion for evaluating GMOs containing pesticides, as they wended their way to commercial authorization. In 2006, I watched Fellous speak in front of Paul Moreira's camera,[1] uncomfortable in his chair, crossing and uncrossing his legs, stammering to make his point. He was defending Monsanto's mere three-month tests on rats fed GMOs. He said these tests were sufficient for him to authorize a certain number of GMOs in Europe, despite proven side effects to the kidneys. From his position as France's chief expert—according to the reporter—he maintained that he had no doubt as to their innocuity. As he put it: "We'll never see a toxic GMO. There is no question about it. After four or five years, maybe . . . "—in other words, after two to three times the life expectancy of the rats! And he was the "number one" expert?

Despite his unconvincing TV appearance, he wrote to France 5 (*Le Magazine de la santé*—a health show on TV) in 2010, and then to the CSA[2] in outrage over the airtime they granted me to discuss our research. This early research focused on Monsanto's data that downplayed the toxic effects of its Roundup-tolerant GMOs—he same ones that Fellous had endorsed and approved for authorization. In the letter, he lashed out at me. He said I was "a subversive activist, anti-GMO militant, sower of fear, disputed and deemed extremely controversial by the scientific community." The "scientific community" was made up of the various academies, agencies, and the French Association of Biotechnologies (an association he presided and in whose name he was writing). On January 18, 2011, in the first of the seven lawsuits I filed, the 17th Correctional Chamber of the High Court of Paris ruled this very association civically responsible, along with Fellous, who was accused of defamation towards a civil servant in the exercise of his functions, "in the name of the French people."

As early as the next day, Fellous gave an account of his loss to Monsanto, omitting a few of the details of his conviction, and assuring them that their common battle for "green bio-technology" would continue with the company's "invaluable support" (MONGLY02997262)! But until we discovered the Monsanto Papers, we didn't know about the direct links between Fellous and Monsanto. As Gaël Lombart of the newspaper *Le Parisien*[3] put it, "The Frenchman Marc Fellous, a signatory of the letter [to *FoodChemTox* calling for the retraction of our study], is a member of *AgBioChatter*. He claimed to have no knowledge of Monsanto's machinations, or of the company's participation in *AgBioChatter*. 'I didn't know,' he said. 'In any case, I have no contact with Monsanto.'"

In reality, Monsanto—notably Eric Sachs—knew the *Association Française des Biotechnologies Végétales* well (AFBV—the French Association of Plant Biotechnologies) and a certain number of its members, including Fellous, and they all also knew one another.

On January 26, 2011, Donna Farmer worried that Monsanto's name appeared in direct connection to a defamation campaign launched by the AFBV as early as 2009 against one of my articles on the effect of Roundup on human cells. She cited Claude Allègre, Axel Kahn, and in the lead, Marc Fellous, the very three I had mentioned in court (MONGLY02428282). Claude Allègre, a former Minister of Research, wrote libelous things about me very soon after the publication of our 2012 study, in the magazine *La Provence*.[4] He had, among other things, made it possible for Axel Kahn to land a project with Rhône-Poulenc who—as revealed by the press—put him on their payroll. To be precise, it was a massive project with Limagrain, called "Genoplante" that was endowed with a budget of some 1.4 billion Francs—a huge sum for the time—and of course financed by taxpayers. Around 1998 or 1999, Axel Kahn left the commission for GMOs (Commission du Génie Biomoléculaire/CGB) and was replaced by Marc Fellous. The very same Fellous who wasn't averse to lying: He went as far as producing a document for the court[5] with our colleague's[6] false signature as evidence to support his reasoning. He was not the only liar in the network.

The AgBioChatter forum was where all of this high society came together to plot and strategize under the iron fisted rule of Monsanto's own Eric Sachs. Kevin Folta, Professor and Chairman of the Horticultural Sciences Department at the University of Florida, belonged to the forum. Folta also wrote and produced podcasts in another forum, *Talking Biotech*, under pseudonyms, to promote pesticides and GMOs. In 2015 he used the forum to discredit, among others, any inconvenient findings of the OMS and IARC on glyphosate.

In October 2017, he was asked how much money the biotechnology sector had given to the horticultural department at the

University of Florida. Folta responded that he had gone through all the records and found no donations in the past five years. Nonetheless, the university[7] received $12 million between 2013 and 2014, and the donors are listed as follows: Monsanto is listed in the golden tier (gifts of at least $1 million), Syngenta in the diamond tier (gifts of at least $10 million), followed by BASF ($1 million), and Pioneer Hi-Bred (at least $100,000). Folta suggested to Monsanto that they hide the money from the public eye, so that no conflicts of interest would be apparent. He later found himself on the wrong side of the law, questioned in Illinois over his actions in favor of Monsanto and his work as a consultant for Bayer earning $600 an hour for twenty-five hours in 2018.[8]

But back to Marc Fellous and the AFBV. The ruling against them came about because they declared that my work was being supported by Greenpeace. It was true, up to a point, but it only concerned a sum of around ten thousand Euros, or 3.45 percent of our research funds over several years, as was demonstrated by an accountant during the court case. That sum was spent on a scholarship to allow a student in charge of deciphering Monsanto's toxicological raw data on GMOs to attend a statistics course for a few months, a cost not covered by the government. The court must have had to bite their tongue when the AFBV and their president Marc Fellous twice claimed their "total independence."

Meanwhile, a quick look at the list my lawyer Maître Bernard Dartevelle came up with, showing the titles and activities of AFBV's members, says it all. Three of the association's founding members worked at this time for Limagrain, the large French seed company; one worked for the French Union of Seed Companies; two others for Maïz'Europ'; another for Maïsadour (with a turnover of 1 billion Euros); one for Rhône-Poulenc Agro, or Aventis Céréáles; and last but not least, the list included the CEO and the President of Champagne Céréales (with a turnover of 840 million). The list also includes members who work for Agralys and Euralys, and it goes on. Thus, there may have been conflicts of interest for Marc Fellous' so-called "independent" association that cloaked itself in "official science." Rotten science?

Axel Kahn was one of the sponsors of the association. He was formerly president of CGB, the outfit that authorized Novartis' first transgenic corn containing an antibiotic-resistant gene. Claude Allègre, the infamous geochemist and former minister, was another.

Perhaps you'd like to know more about the AFBV's members and their patents. Roland Douce, a member of the American Academy of Science worked for Nestlé on transgenic coffee and also worked for Rhône-Poulenc. Lise Jouanin worked with Imperial Chemical Industries, Genoplante, Limagrain and Bayer; Alain Toppan, a creator of GMOs who regularly solicited authorizations from the CGB, turned up at Sanofi Elf and Elf Aquitaine. We almost forgot Marc Van Montagu, the decorated inventor of GMOs in 1983, who worked with Plant Genetic Systems and then Bayer AG. One of his students was hired by the European Food Safety Authority and attacked our work. To wrap up this non-exhaustive list, let us also mention Marcel Kuntz who along with Syngenta, applied for two patents for transgenic plants.

An almost entirely French group of creators of GMOs containing pesticide and petroleum products. And they call themselves independent. The only thing they are independent from is not the industry, but more likely the general population.

Let's talk about Marcel Kuntz for a moment, the author of *The Seralini Affair: A Militant Science Reaches a Dead End*, a 2019 pamphlet distributed to lawmakers in several languages. It has the golden stamp of the Agricultural Academy of France and reassures the reader that the publication has "no source of revenue linked to the commercialization of agricultural products, the biotechnology or the agrochemical industry." Kuntz held patents on techniques of genetic transformation, a livelihood connected to GMOs, if only distantly, without a doubt. He nonetheless did his best to belittle our research with countertruths, with the unwavering support of his colleagues. Kuntz was also a member of *AgBioChatter* and espoused their arguments. According to him we got it all wrong, and we had too much influence, especially since we appeared on the show *Envoyé spécial* (Correspondent). This was around the time we were advising the victorious American lawyers in the lawsuits against Monsanto.

The images of the rat tumors from our 2012 study, broadcast during that same show, still ate away at our opponents, most likely causing them to suffer even more than the rats themselves. I was put on their newly created blacklist. When their blacklist went public, Bayer minimized the whole thing. The files Monsanto had collected on us with the help of a PR agency were illegal. I registered a complaint against the company in 2020, as did the other people affected.

Again, during the 2011 lawsuit, Marc Fellous stood his ground and explained via his lawyer that "our country's top scientists" had called our work into question. He added an impressive petition from the members of the French Academy of Agriculture to the file, (this was contrary to the Academy's own bylaws). To defend the idea of this petition, Fellous produced the fake signature of a top scientist. He also quoted select highly placed officials, making them appear to be independent authorities. In fact, they were a small congregation of like-minded industry supporters, members of the academy of medicine, the academy of agriculture, the academy of science, the AFSSA (*L'Agence française de sécurité sanitaire des aliments*—the French national authority on food safety), and the HCB (*Haut Conseil des biotechnologies*—the high council on biotechnology—the new incarnation of the CGB which was dissolved in 2008 following our disagreements).

They introduced and influenced one another and schemed, as was demonstrated in an investigation by *Lyon Capitale*[9] and the interview with the academician (Academy member in our defense) and toxicology mathematician-statistician Paul Deheuvels.[10] Deheuvels had never seen the petition before its release, but it bore a strange resemblance to Monsanto's argument against us, signed by its henchmen just after our 2012 publication. Perhaps it's not surprising, given their usual way of operating. Professor Deheuvels testified in favor of the quality of my work during the lawsuits I filed. He explained that the president of the Academy of Science, Jean-François Bach—who had registered patents with the National Institute of Industrial Property (INPI), notably with Rhône-Poulenc and Sanofi—had formed his opinion with twelve chosen members out of 2,592 registered (two for each academy), the elite of French academia. This was good fodder for a statistician: The French elite in a nutshell.

The controversy this group stirred up continued into 2012, despite our winning the 2011 court case. Three people in particular played an active part in disseminating the academies' announcement to PR agencies: Georges Pelletier, of the Academy of Agriculture, the CGB, and a patent holder on GMOs; Roland Douce of the group we defeated, the French Association of Plant Biotechnologies (AFBV), and from the Academy; and Gérard Pascal, of the AFBV, the International Life Sciences Institute, and the Science Media Center. They even had the announcement relatively rapidly translated in order to denigrate our work in the face of the campaign in California to label GMOs.

The GMOs Could Have Been Labeled in the USA

In October 2012 Henry Miller campaigned in the name of Stanford University (who promptly condemned his outrageous claim) against the proposal to label GMOs and distributed leaflets restating that our studies were "fraudulent," as in his *Forbes* piece that the magazine finally pulled—but only in 2017—because of conflict of interest.

In 2012, there were demonstrations in the streets of California led by concerned citizens ready to fight in support of our work, to denounce the toxicity of GMOs, and to support their labeling. A doctor told me the story of how she had marched with her colleagues, naked like my rodents except for sandwich boards showing the images of the tumor-infested rats fed the Roundup-tolerant GMOs (the ones authorized by Marc Fellous to be exact). Our position on labeling followed our clarification of the risks that could have been taken into account in order to change the situation in the United States—all this to give people a choice. Monsanto, meanwhile, with the help of Miller and the French academies, had directed all its energy against my studies. They were awoken by their lobbyists. The proposal to label GMOs lost by a hair in the California vote (53 percent against). The companies that illegally financed the campaign would again be condemned in Court in 2020.[11]

The Illegal Collecting of Information Goes On

Around the time Bayer began the takeover of Monsanto in 2019, the law firm Sidley Austin LLP reached out to the people Monsanto had been keeping illegal records on over several years. The German company was worried about its reputation (though it wasn't foreign to record-keeping, given its wartime activities). A group of us, composed of researchers (myself and Robert Bellé included), reporters (*Le Monde*), and politicians (among them Ségolène Royal), had filed a civil suit against Monsanto-Bayer for collecting these records. They basically tracked an individual's stance on GMOs, etc., and allowed the company to sort out who they could recruit and who they should exclude from their circle.

The list looked a lot like something that might be drawn up by a totalitarian government. Sidley approached the people on the list with an apology from Bayer. It was utterly ridiculous. In my case, Sidley sent a file via registered mail (on elegant, embossed paper!) containing the "confidential information" that had been collected on me, this in the name of transparency. In it was a one-page slide presentation with the image of a rapeseed field, entitled, "Glyphosate in France," and a blacked-out square that must have shown my name and the name of the author of the investigation. They supposedly did that for privacy. There was also another page showing an array of black squares. Bayer was obviously continuing to mock the researchers. I filed a lawsuit in early September 2019, with the help of the William Bourdon law firm. At the same time, I found out from the organizers that Bayer had sprung into action—threats and all—to try to prevent me from opening as a speaker for the conference "For a Pesticide-free Africa" in Kenya that summer.

The Networks Still Work

In 2019, Monsanto-Bayer, the "spider-octopus," continued its bribery schemes as if the Monsanto Papers had never existed. Bayer promised to clean up Monsanto's vile practices, but in order to do so, they hired the company's own infamous toxicologist Donna Farmer to head the effort (the head of the toxicology department). When Bayer took over Monsanto, they were faced with 48,600 lawsuits in the United States for covering up the health risks of Roundup. In 2020 they began to amicably settle class action lawsuits in order to avoid trials.[12] These very companies had now known of the dangers of their products for over fifty years, but continued, under wraps, almost in broad daylight, to sell the banned pesticides to poor countries or those with lax regulations. Even with the threat of lawsuits, they made more money this way.[13]

On February 25, 2019, Monsanto-Bayer's[14] lawyers gave the judge a long list of topics they didn't want broached during the trials: no mention of other lawsuits against Monsanto; no evidence surrounding the firm's public relations efforts; no comparisons with the tobacco industry; no association made with any of the company's other controversial products (Agent Orange or PCB); no information about the company's wealth; and, the most mind-blowing one, no talk about Bayer's role in World War II.

The authors of the company's slogan "Bayer, Science for a Better Life," must have been trying to pass for cherubs. Their parent company manufactured Zyklon B[15] for Hitler's concentration camps—of which they built and administered at least one (Auschwitz III)—even as they paid dividends to shareholders, among them Bayer,[16] without any compensation to the victims. This went on for decades and was never completely resolved.

The people quoted in this chapter (the network of companies, law firms, and individuals in favor of pesticide-rich GMOs) have used the media to quash the whistleblowers. They were experts or members of official public health agencies who mutually supported one another. Gérard Pascal, the toxicologist from the National Institute of Agricultural Research, and the CGB, who testified for Fellous, had been the first one to speak up in the press against my 2012 study. He was proud to say he participated in the selection of experts on GMOs from the EFSA (and ILSI). He had worked for Danone and Nestlé. In court he left out his role as an official lobbyist for ILSI (you could see his CV on the site ilsi.edu.[17]) Apparently, this was enough to grant him the impartiality he needed for the job.

We detailed all of these connections, and many more, in our book *Hidden Poisons or Culinary Delights*, as did the newspaper *Le Monde*, and investigators such as Benjamin Sourice.[18] The scandals piled up, heads of agencies were fired, but the core group of experts is still there, lulled by the music of money. Kathryn Forgie, who revealed the Monsanto Papers, spoke about them in great detail at the European parliament. We don't feel it necessary to add any more details (phew! sighs my coauthor, brandishing a drink, half-dead from despair in the face of such an avalanche of details).

Chapter 13
TOXIC NEWS: THE INFILTRATION OF THE MEDIA

First instance lawsuits and appeals against the magazine
Marianne, and against a grumpy retiree. The misdeeds of the
too rarely condemned appointed bloggers who spread defamation
on social media.

There are now two losers condemned by the French justice system for their defamation of our work and persons: Marc Fellous, member of AgBioChatter, and the Association of Plant Biotechnologies (AFBV), over which he presides. Together they represent a "French appendix" to the premier network of worldwide lobbies. My coalition of supporters, my own team, and I myself, separately won these two lawsuits against lobbies linked to Monsanto. Even if it wasn't always obvious, Fellous wore two hats. He played a key role for the government when field tests of agricultural GMOs were approved, which helped him win the authorizations to commercialize the ones that he supported. And thanks to his role as president of the Biomolecular Engineering Commission, these authorizations later extended to all of Europe. His predecessor, the geneticist Axel Kahn, had already made France the European leader in agricultural GMO experimentation a few years earlier. A reminder that these people specialized in human genetics and had no competency in plant biology.

Next to appear were the two, two-time losers (the magazine *Marianne*, and the journalist Jean-Claude Jaillette), both in their trials and on appeal. You will see how they connect to the first two. This covers six out of the seven trials.

Together, this group was trying to unleash a backlash against our research. Their mission was to make people believe that a real and diverse "international community," was working against us, brought together by science, when in fact all that united it was dollar bills and the fact that its members could all be traced back to the sugar-daddy Monsanto. In other words, they had the means. It meant that we too would have to work the system, but how?

When *Marianne* Plagiarized *Forbes*

Let's take the same group and start over. In 1997, Jean-François Kahn, Axel Kahn's brother, founded the weekly news magazine *Marianne*. The magazine loved a good scoop. He remained at its helm for ten years, but continued to write for them up until 2011. He collaborated with his brother on the book *Comme deux frères* (*Just Like Two Brothers*).[1] In 2009 his brother Axel wrote the preface for a book by the *Marianne* reporter Jean-Claude Jaillette, in which he didn't hide his position. The title of the book is an exhortation: *Sauvez les OGM* (*Save the GMOs*).[2] Here's a quote: "To pass over the potential of GMOs would be suicidal, if not criminal." So it was no great surprise when Jaillette's statements were ruled defamatory by a court. And not only were they defamatory but they happened to be lifted directly from the writings of one of the foremost pro-GMO lobbyists, the one who drove fast to escape his own lies—do you remember him?

The Biomolecular Engineering Commission's members—past and present—felt they were being called into question in the face of our studies on the toxicity of the genetically modified corn GMO NK 603, bred to absorb Roundup. They had contributed to its authorization. If they had given the green light to a toxic product, they could be held legally responsible. I was on the commission at the time and had clashed vehemently over the slipshod three-month-long tests Monsanto had performed. But this core group of compromised people validated the positive views that upheld the innocuity of the barely tested GMO. My team had gone on to discover that in their second year of life, rats that ate food contaminated by this GMO—or by Roundup—developed very serious illnesses.

A quick reminder: Monsanto focused on sweeping the effects of Roundup under the rug in remote-controlled debates, just as Donna Farmer and AgBioChatter had instructed. Of course, one should only discuss the images of rats' tumors that went viral—poor things—but not the deadly effects on the animals' kidneys and liver, something we would later confirm in leading scientific journals. To us, it was a dishonest distortion of the debate—a debate into which a highly visible media personality and mathematician threw himself.

In 2012, in front of the National Assembly, Cédric Villani declared that he didn't believe our results, not because he considered himself a specialist in the field, but because "his friends had told him so." These "friends" all belonged to the Academy of Science. Since that time, fear not, his political ambitions have directed him to ecology.

The members of the Commission were often joined by the Association of Plant Biotechnologies (you'll recall the two hats worn by their president), the most militant supporters of GMOs, were among the first forty signatures of the vehement arguments against us.[3] But that was just their opinion. Their group was comprised of: Marc Fellous, Gérard Pascal, Georges Pelletier (of the government Commission, as well as the activist association, and one of a select group of the agitators against my work from the Academy of Science). Other signatories included Yvette Dattée and Francine Casse, who served along with Gérard Pascal on Axel

Kahn's commission, and were already testifying on behalf of the industry in the case against the looters of GMO crops. We also would find Francis Quetier (Genoplante), Marc Van Montagu, and, what a surprise, Wayne Parrot of AgBioChatter. They would be joined the following week by Bruce Chassy, and Ingo Potrykus (who detained multiple patents on the GMO for golden rice). The entire Monsanto intelligentsia was there in all its glory, passing for an independent community—the "official science."

Fellous requested an audience with the Superior Counsel on Audiovisual to badmouth us. We ourselves granted him this request, so that people would be able to appreciate his competence as "chief French expert." He impelled *Marianne* to publish it. It is a citation from MONGLY02539190, in an email dated September 19, 2012, in which he writes to Monsanto in reaction to our study, just published online: "I am speaking to journalists today and tomorrow. From what I understood from the GES paper is the effect of Roundup itself on rats. In the water, in MONK603[4] or in GMO maize alone. The article is on the effect of the herbicide by itself. What do we know on the acute and sub-chronic assay of the Roundup, and its adjuvant? Do the effect on rat (ant [sic] other animals models) found such tumours and lethality, I doubt? Do there [sic] data? Marc." You can tell from the bad English that this was written in a state of panic. This message in a bottle shows not only his shortcomings—besides spelling and grammar—but also his blind confidence in the parent company. The very same day Eric Sachs moved to reassure him, as he summoned the troupes from AgBioChatter and the other heads of Monsanto: "It would be good to provide arguments for Marc."

Fellous's colleagues got together and sent a letter against me to Jean-Claude Jaillette for him to publish in *Marianne*. What a surprise! They stood up for the friends of Danone, and one member in particular who was trying to defend Monsanto against us in the eyes of the French government.

In all the signatories of the letter, there was more than one industry supporter in disguise. On September 21, 2012, the Monsanto toxicologist Bruce Hammond sent the group's condemnatory comments, along with those of all the companies called to the rescue (Bayer, BASF, Syngenta, Pioneer (MONGLY01126903)) to a certain Andrew Bartholomaeus. He was with the Australian and New Zealander regulatory authorities. Bartholomaeus, a former member of the International Life Sciences Institute, and an expert for the Food and Agricultural Organization of the UN and for WHO, had first published in the same scientific journal where our study appeared, *Food and Chemical Toxicology*. He had also worked on deregulating nanoparticles in food and agriculture (there are many, derived from heavy metals, in the formulants of pesticides. In 2020 the regulatory agencies were worried about their use, a transparency the agencies themselves did not insist upon).[5] A majority of industries usually in competition with one another came together in order to denigrate our work by pseudo-specialists. They all had a bone to pick with us as our research called into question their right to market their products.

The op-ed in *Marianne*, signed J.-C. J., was printed like an epitaph, with a black border. During the trial Jean-Claude Jaillette blamed the editor in chief: They were not happy that their network was revealed.

Marianne and Jaillette defended one another to the bitter end. There was no doubt as to the aim of the angry words outlined in black—the ones we attacked in court. J.-C.J. never interviewed me, giving me the chance to respond, and abstained from mentioning the hundreds of scientists from around the world who vouched for our work. My lawyer, maître Bernard Dartevelle, highlighted their testimonies in the trial. Those scientists not only supported us but disapproved the retraction of our study, which they saw as scientific censorship.[6]

J.-C. J. based his arguments solely on the forty signatories, some of whom are quoted above, while others remained unknown. "They expressed their skepticism about the scientific process, as well as their anger." His words were harsh: "communication operation," "scientific fraud in which the methodology supports the results written before the experiment was performed," etc.

During the trial it was revealed that up until the *Marianne* piece, no other article had mentioned fraud other than the one by the Monsanto's famous honorary author Henry Miller, together with his collaborator Bruce Chassy. Jaillette admitted to the court that he hadn't cited his sources for this serious accusation. One of the most critical voices of them all was Gérard Pascal, with whom Jaillette had lunched the day before the article appeared; he admitted he knew Pascal was a member of the International Life Sciences Institute network.

In its ruling, the High Court of Paris fined *Marianne* and Jaillette. The two dragged their feet over paying their fines, damages, and interest. The High Court made clear in the trial and in the appeal (at the end of 2015), that one article alone, published in *Forbes* magazine on the 24th of September 2012, expressly accused us of "attempted fraud"[7] and also that the "researchers' goal was to obtain a false and pre-established result. The study was conceived precisely to obtain a false result [. . .] culminating in the appearance of large and grotesque tumors." The judgement went further to emphasize that "its authors, Henry Miller and Bruce Chassy, were known for their ties to big industry, in particular Monsanto." That said it all. The *Marianne* piece was solely based on the piece in *Forbes*. Jaillette and *Marianne* were kowtowing to Monsanto by smoking out the forty signatories. Two years following this ruling, Miller and Chassy's articles were retracted by *Forbes* for conflicts of interest.

Let's rewind for a moment. In October 2012, the same group, but with more Americans added to it, wrote to the journal that published me—as always, concealing their links to Monsanto—requesting the retraction of the study. Among them were Chassy, Miller, and Fellous, and also Selim Cetiner, who expressed his doubts as to the validity of the arguments. Before they sent it, they gave their version to Eric Sachs at Monsanto (MONGLY06990942).

Beyond the Bosphorus, Selim Comes Face to Face with Fake Science

Monsanto takes great care in how it presents its fake science around the world, and tries to persuade others to believe in it. To this end the company founded a Yahoo discussion group, AgBioChatter, under the aegis of its minions, the academics paid to write what Monsanto wanted to hear. The question is what was their initial goal? "To discuss Mr. Seralini's controversial results," says one of the members.[8] So it was on AgBioChatter that one of its two moderators, a coauthor of Henry Miller's, provided talking points and arguments that would be used the world over by the likeminded in order to counter Seralini's study, calling it "fraudulent."[9]

AgBioChatter is in fact a private message board used by the agrochemical industry and its allies to coordinate their lobbying. Nothing said there can be directly quoted in public. It was here that Monsanto's arguments were prepared and formulated. It listed as its members academics, members of the senior staff of the industry, and upper-echelon public relations figures. Of particular note were the pro-GMO academics, who used it to share arguments to be used against studies disturbing for them.

This is where we meet Selim Cetiner, of the Sabanci University of Istanbul. He was recruited by Monsanto as a consultant on the security of biotechnologies. He found himself perplexed by the inconsistency of Monsanto's arguments (MONGLY-00900531). For example, the most distorted criticism the company put out bothered the brilliant Turk: "The rat used was an inappropriate type," but the same type was used by Monsanto, meaning that the authorizations should have been retracted. All it took was a heated response from Erich Sachs to stop him from asking his pertinent question, put him back in his place, and get him to sign the letter to the journal asking for the retraction.

Since December 2008 at least (MONGLY01192418) and in the same way, we found ready-made answers, in keeping with the company's stance, aimed at consolidating their viewpoint around the world, especially against my studies. They can be found in files stamped "Company confidential for internal use only," sounding the alarm when a study arises that deals with the toxicity of Roundup, with a bold-faced warning at the bottom of the page, iterating the official version to be used by the health agencies—supposedly the Truth emanating from the scientific "community."

As early as January 19, 2011, Daniel Goldstein (Dan) had written to, among others, Eric Sachs to say that on the subject of the "Seralini vs. Fellous defamation case" (MONGLY02997262) that the libel cases are a "recipe for disaster." Monsanto was afraid of us. It's true that among all the known cases of whistleblowers, it is in the Seralini case and only in that one, that you will find a lone researcher attacking the lobbies, with the support of the team. In all other instances, it's the corporation who attacks first.

Monsanto is afraid of the truth. That's why the company hides behind its minions, who are both pseudo-scientific in their arguments and pseudo-industrial in their financial compensations. It's more than sad: it's nauseating.

Who Is the Puppet?
Out of all of today's forms of media, social media is more widely read than many newspapers, and goes further in its attacks. It wasn't yet encumbered by lawsuits, and at the beginning felt the growing wings of absolute freedom due to an unhealthy anonymity and the cover of easily accessible pseudonyms. The blogosphere has taken off, often without any journalistic analysis: Tweets, messages

and websites sometimes give way to hatred and serve as a spillway for frustrations.

On October 25, 2012, an anonymous story was published on the blog *La Lettre du Cotentin/The Cotentin Newsletter*[10] entitled "Seralini the Jester," criticizing the methodology, goal, and results of our study of September 19. The story didn't make any new points, but more or less used the same talking points as Monsanto. On October 26, 2012, it was picked up by *ForumPhyto*, the site for pesticide sellers—a major union between farmers and industry. They kept it online, even though it had been retracted by the original website and it was still accessible as late as the beginning of 2020. The story was a reminder of the value of pesticides, and the danger, in terms of famines, poverty and war, of a setback in their use—the usual distortions.

The story depicts Seralini as "a dangerous and mad scientific buffoon, a rascal of science. [. . .] An academic who abuses the power of his white lab coat to deceive the public, [. . .] a Puppet from the University of Caen," employing "special effects," "deliberate lies," "scientific dishonesty," "major scientific fraud," "concealment," "propagation of fake news," and "squandering" of public funds. We should "remove him from office" and proceed to his "imprisonment."

Following our complaint, the delinquency suppression squad identified the author of those lines: Dubost, Daniel Emile Louis, born on November 12, 1938, to Aristide Dubost and Denise Postel, retiree without a record from . . . the University of Caen, living in Reville in the Manche region. Could all this vehemence be coming from a stockholder in the pesticide industry? Or would it rather be a jealous and frustrated colleague who published little to nothing while at the University of Caen? If we will never know this person's precise motivations, investigators were able to reveal the ties that other bloggers or trolls had with the lobbies. In this case, the scientific academician Yvon Le Maho testified in my favor as to the validity of my work and his objection to the villainy of these attacks.

The university where I worked and my hierarchy received vengeful phone calls from the perpetrator slandering me, or from the people who liked his message. In our world of internal competition for job openings in a team, for recognition, teaching posts, or promotions, the jealous play the wise guy and feed on controversy in order to bury a competitor.

On November 18, 2016, after two hearing requests, Daniel Dubost was charged with defamation of a civil servant in the exercise of his duties.

Others were better organized and had financial backing from the industry, like pesticide manufacturers, or even freelance public relations firms. These actors had more influence and endurance. Their aim was to throw a wrench into the system in order to prevent our research from being readily available. They wanted to prevent a ban on their products by making our studies seem unreliable, or by attacking me personally. They tended to attack ecology itself, or the reporters who exposed the industry, such as Paul Moreira or Élise Lucet.

Of these many perpetrators from around the globe, here are two of the "stars" who wouldn't give up. They were intent on preserving their anonymity. The first one, Wackes Seppi, whose real name is André Heitz, is a retired UN functionary who used to work on patents and intellectual property. His job was to go to Brussels where all the lobbies hung out, to reinforce ties with large corporations. The second one is Stéphane Adrover, also known as Anton Suwalki, a statistician from the National Institute of Statistics and Economic Studies. He too had time on his hands but, reading his CV, no competence in biology, agriculture, or journalism.[11] Between 2008 and 2012, he was on the board of the French Association for Scientific Information (AFIS), whose president was Louis-Marie Houdebine. This gentleman, who was a geneticist for the National

Institute of Agricultural Research (INRA), created genetically modified fluorescent rabbits in his laboratory. At least one of them (a "fluo rabbit") found itself a home with an internet-surfing artist who is one of the most vocal proponents of GMOs under the sun.

Other members of AFIS include Fellous, Pascal, Datée, Kuntz, Joudrier—the very ones who promoted the authorization of pesticide-tolerant GMOs in Europe. This same association contributed the information that corrupted my Wikipedia profile as it became available, full of misinformation, in every language, around the world. Adrover and Heitz were convicted under their real names of defamation on July 1, 2019 in the case of Paul Moreira, the director of the documentary "*Les OGM bientôt dans nos assiettes*" or "*GMOs coming soon to your plate*." They had used their usual slander, including "docu-liar," "incompetent," "shoddy." and "juvenile."

To complete the picture, another sad example was Gil Rivière-Wekstein (he didn't resort to pseudonyms), who blatantly posted Monsanto's arguments against our work on his website Agriculture and Environment (*Agriculture & Environnement*). But it didn't stop there. The AFIS liked his work and gave his book a nice review. He didn't hide his name, only his *larouchist* deference as the reporter Fabrice Nicolino called it (in brief, fans of dictators). He shot down anyone who stood up against GMOs or pesticides, or who criticized the chemical industry. He was also a proponent of the return of DDT, something he expanded upon on the website "*Alerte environnement*." He disguises his blogs with eco-friendly sounding titles, in an effort to entice impressionable readers. This industry "publisher" was convicted of libel against another civil servant in 2007 and 2009. Despite this, he continues his poisonous enterprises just like the rest of them. He apparently coined the term "agri-bashing"[12] on Twitter as early as 2016. At the time he was the head of a micro-consulting firm that made its money (250,000 Euros in turnover) printing for the pesticide industry.

The concept of "agri-bashing" was picked up with a vengeance by the National Federation of Agricultural Holders' Unions (FNSEA) to attack anti-GMO and "organic" groups. Mainly it was used to convince the French government to establish a police unit to watch and suppress any attack on intensive farming. Strangely the unit was named "Demeter," the goddess of the harvest, and also the name of a quality seal in biodynamic organic farming. They are all part of the network of slander interconnected with the government that sows confusion in order to stay in power.

Chapter 14

A SOCIETY WITHOUT ROTTEN SCIENCE

Never in the history of humanity has science so closely approximated religion as it does today, what with its dogmas, its critical judgement mutated by the technocracy, and its industrial-power leadership. Leaders and the press consult the pundits in white coats as if they were the clergy.

This was never in greater evidence than during the Covid-19 pandemic that began in 2019–2020,[1] when world leaders leaned on appointed physicians to justify exceptional lockdown measures. Equally shocking was to see public health agencies geared up to search for side effects of a very common drug used for decades, that empirically proved its efficacy while the authorities pay less attention to the safety tests of vaccines; juicy for industry, but not subjected to the same decades-long safety reviews. In Washington state, in 2019, Robert F. Kennedy Jr. led a case against the US government for deceiving the public about the efficacy and safety of vaccines for over thirty years. One can hope that this case will have international legal repercussions on parents' freedom of choice in an era when mass-vaccination is touted as the only solution to curbing the Covid-19 disease.

Promoting this idea, is an old Monsanto ally, Bill Gates—also an ally of the international pharmaceutical industry, of GMO biotechnology, and of information technology.[2] He has been promoting mass vaccination through his foundation and collecting people's health data in Google databases. The Bill & Melinda Gates Foundation has been a shareholder in Monsanto since 2010 and is one of the major funders of the WHO. Let's imagine for a minute that a vaccine based on GMOs, for example, made from tobacco,[3] has every chance of being introduced as the savior of humanity in the face of Covid-19. Bill will be counting on the medical authorities. Over several decades post-WWII, public health agencies, environmental agencies, and the academies, be they national, continental like the EFSA or EPA, or international like the WHO, have been infiltrated by industry lobbyists, as have the large research organizations. They too can count on sponsors like Gates.

The dogma of international finance, like a religion, has slowly taken charge. It has taken over the most fundamental principles of toxicology, of medicine and its founding principles, of genetics, and biodiversity, to give a few examples. The basic scientific fundamentals of the observation of detail have been lost in reductionism and private property. This is the case for patents on DNA sequences, transgenic plants, and information networks.

Never before has science behaved in such a criminal and blind way in the face of damages to our health and to the earth. The commissions put in place way too late have made observations, but it is the scientists who justified the mass marketing without adequate testing of plastics, pesticides, and detergents corrosive to the brain that exist in tobacco, GMOs containing pesticides, toxic nanoparticles, shoddy vaccines that have either been badly made or barely tested, hormonal and nerve disruptors, to give some examples.

From my vantage point on the commissions, I agreed to serve on for nine years, I watched government and international commissions laugh and propagate their dishonest myths by organizing a campaign of disinformation. They knew. I believed that the decision makers at multinational corporations who marketed these poisons were just selfish, but in the examples I observed firsthand, a criminal element actually reigned. There is as much unethical behavior in science and medicine as in any other field. We upheld our belief that these fields were above that, despite the cathedrals of commerce, but found this was not the case.

Those who play the multinational game can never simultaneously labor for the common good. Toxic jobs and benefits exist, and those who want to participate in these activities should come clean: Their money has smell, taste, and deleterious side effects.

Their corruption rests for the most part on the illegal secretive health tests performed on products that are taking over our planet and our bodies. Their purposefully biased interpretations of the effects and side effects of these products are of a criminal nature. The absence of an open dissenting expertise, such as there is in a democratic court of justice, reinforces a techno-scientific dictatorship. Without adequate measures to ensure the necessary changes, and without the means to ensure a common future for society, humanity will be irretrievably damaged, and a vast part of the ecosystem with it.

We will continue our work to ensure that this will not happen.

Endnotes

Foreword

1. This became official in 1954.
2. Punjab was adversely affected by the actions of the separatists and the repression that followed. The regime became totalitarian, and misery spread.
3. Charles Benbrook, "Trends in glyphosate herbicide use in the United States and globally," *ResearchGate*, February 2016, www.researchgate.net/publication/292944439.
4. www.baumhedlundlaw.com/toxic-tort-law/monsanto-roundup -lawsuit/where-is-glyphosatebanned/ and www. carlsonattorneys.com/news-and-update/banning-roundup.
5. Kavin Senapathy, "Opinion: I Was Lured into Monsanto's GMO Crusade. Here's What I Learned," *UnDark*, June 27, 2019, undark.org/2019/06/27/monsanto-gmo-crusade/.
6. https://en.monsantotribunal.org/.
7. Camila Domonoske, "Monsanto No More: Agri-Chemical Giant's Name Dropped in Bayer Acquisition," NPR, June 4, 2018, www.npr.org/sections/ thetwo-way/2018/06/04/616772911 /monsanto-no-more-agri-chemical-giants-name-dropped-in -bayer-acquisition.

The Seralini Affair and Monsanto's Manipulation of Science

1. www.baumhedlundlaw.com/toxic-tort-law/monsanto-roundup -lawsuit/monsanto-secret-documents/.

2. Monsanto. MONGLY10145298. Business Performance, David Saltmiras, 20 August 2013. Available from: http://baumhedlundlaw.com/pdf/monsanto-documents/8-Monsanto-Scientist-Admits-to-Leveraging-Relationship-with-Food-and-Chemical-Toxicology-Journal.pdf.

Monsanto vs. Seralini: Pulling Back the Curtain
1. https://www.iarc.who.int/news-events/.glyphosate-monograph-now-available/.
2. https://www.elsevier.com/about/press-releases/research-and-journals/elsevier-announces-article-retraction-from-journal-food-and-chemical-toxicology.
3. https://usrtk.org/wp-content/uploads/2019/08/Eva-Novotny-paper-on-corruption-behind-Seralini-retraction.pdf.
4. Ibid.
5. https://pubmed.ncbi.nlm.nih.gov/22999595/.
6. Ibid.
7. https://www.baumhedlundlaw.com/documents/pdf/monsanto-documents-2/email-re-freedom-to-operate-and-stewardship_redacted.pdf.
8. https://www.baumhedlundlaw.com/documents/pdf/monsanto-documents/27-internal-monsanto-email-you-cannot-say-that-roundup-is-not-a-carcinogen.pdf.
9. https://www.baumhedlundlaw.com/documents/pdf/monsanto-documents-2/MONGLY14441101-MONGLY14441108.pdf.
10. https://www.huffpost.com/entry/monsantos-spies_n_5d7ba20de4b03b5fc88233c4.
11. https://www.baumhedlundlaw.com/assets/Monsanto%20Roundup%20pages/Secret%20Documents/monsanto-documents-chart-101217.pdf.
12. https://pubmed.ncbi.nlm.nih.gov/20979644/.
13. https://www.baumhedlundlaw.com/documents/pdf/monsanto-documents-2/mongly00985794.pdf.

14. https://www.baumhedlundlaw.com/documents/pdf/monsanto
-documents-2/mongly03081997.pdf.
15. https://usrtk.org/wp-content/uploads/bsk-pdf-manager/2019/03
/Plaintiffs-exhibit-Monsanto-internal-email-string-celebrating
-Seralini-attack.pdf.

Chapter 1

1. Publication date of my team's first study showing the toxicity of Roundup on human cells, much greater than that of the glyphosate it contains: Sophie Richard et al., *Environmental Health Perspectives*, 2005, 113 (6), p. 716–720. Cf. MONGLY00923758, usrtk.org/wp-content/uploads/bsk-pdf -manager/2019/04/Monsanto-2005-criticism-of-Seralini-article .pdf.

Chapter 2

1. Using the keyword "Seralini" for example, you can consult those that were made public, out of two and a half billion, on January 23, 2020: https://www.baumhedlundlaw.com/toxic -tort-law/monsanto-roundup-lawsuit/monsanto-secret -documents/.
2. This is about the product's ingredients, a company secret, along with the details of the tests performed by the company showing its impact on our health and environment.
3. Hervé Kempf, *Reporterre*, October 24, 2017.
4. Stéphane Foucart and Stéphane Horel, *Le Monde*, June 1, 2017 and October 5, 2017; *New York Times*, August 1, 2017; *The Guardian*, September 14, 2017; *La Repubblica*, September 23, 2017.
5. See documentary video *Envoyé Spécial*, (*Investigative Reporter*): "Glyphosate, comment s'en sortir?" (Glyphosate: How Will We Deal With It?). January 17, 2019.

6.   Stéphane Foucart and Stéphane Horel, "L'affaire Seralini ou les coulisses d'un torpillage" ("The Seralini Affair, Or Behind the Scenes of a Torpedoing"), *Le Monde*, October 5, 2017.

7.   Gilles-Éric Seralini et al., *Food and Chemical Toxicology* 50 (2012): 4221–4231.

8.   Eva Novotny, "Retraction by Corruption: the 2012 Seralini Paper," *J. Biol. Phys. Chem.* 18 (2018): 256.

9.   Gilles-Éric Seralini et al., *Environmental Sciences Europe*, 2014, 26, p. 14.

10. Robin Mesnage et al., *Science Reports* 7:39328 (2017); Robin Mesnage et al., *Science Reports* 6:78555 (2010)

11. Report from the Supreme Court of the State of California, August 9, 2018, source US Right to Know (USRTK).

12. See detail in the List of Principal Characters: Bernard Dartevelle's name, my personal lawyer and the lawyer for my team.

13. See "Canadian Lawyers Launch $500M Class-Action Lawsuit against Roundup Makers," sustainablepulse.com/2019/11/22

14. https://usrtk.org/wp-content/uploads/2018/08/Johnson-trial-judy-verdict.pdf.

15. Michael James and Jorge L. Ortiz, "Jury Orders Monsanto to Pay $289 Million to Cancer Patient in Roundup Lawsuit," *USA Today*, August 10, 2018.

16. https://www.youtube.com/watch?v=FYnWPcgEqbQ.

17. MONGLY01314233–270; we will go over these events.

18. usrtk.org/pesticides/.questions-about-epa-monsantocollusion-raised-in-cancer-lawsuits/.

19. "Whack a Mole" and "Let Nothing Go": These were directives the company put into place to fight against detractors.

Chapter 3

1.   *OGM, le vrai débat* (*GMO: The Real Question*), Flammarion, 2000; *Ces OGM qui changent le monde* (*The GMOs That Are*

Changing the World), Flammarion, 2004, rééd. coll. "Champs,"
2010. In 1994 Gilles-Éric Seralini published in books some of
his lectures: *L'Évolution de la matière, de la naissance de l'univers à
l'ADN (The Evolution of Matter: From the Birth of the Universe to
DNA)*, Pocket-Cité des sciences, coll. "Explora," which would
make him a household name. In 1997 his sensibility to ecology
is forshadowed in the book *Le Sursis de l'espèce humaine (A
Reprieve for Humankind)*, a controversial title at the time, that
came out with Belfond; following this publication he became
friends with Jean-Marie Pelt, with whom he later wrote a book.

2. "Génie génétique. Des chercheurs citoyens s'expriment"
    (Genetic Engineering: Citizen Researchers Speak Out), Sang de la
    Terre, 1997.

3. Explained in my book *Tous cobayes (We Are All Guinea pigs)*,
    Flammarion, 2012.

4. Gilles-Éric Seralini et al., *Archives of Environmental
    Contamination and Toxicology* 52 (2007): 593–602. All of the
    articles mentioned and others from that period can be
    downloaded free of charge at www.seralini.fr.

5. According to Monsanto, the negative impacts on the rats'
    health would have to be proportional to the dose and the
    time period in the same way in males as in females in order to
    be considered. As true as that is in short-term toxicology, it is
    stupid in this case. Gilles-Éric Seralini et al., *International
    Journal of Biological Sciences* 5 (2009) 438443.

6. Professor Seralini contributed to professional seminars on
    detoxification at that time organized by Jérôme Douzelet at his
    hotel-restaurant *Le Mas de Rivet (30, Barjac, France)*.

Chapter 4

1. Retracted by Forbes in 2017, following proof that Henry Miller
    was putting his name to Monsanto's articles without declaring
    any conflict of interest, notably on September 26, 2017. He

would do this several times and on various topics: Gary Ruskin, "Henry Miller Dropped by Forbes for Monsanto Ghostwriting Scandal," August 30, 2017, usrtk.org/our -investigations/why-you-cant-trust-henry-miller-on-gmos/.

2. Science Media Center. For further information, read the article by Jonathan Matthews: "Smelling a corporate rat," December 11, 2012, spinwatch.org/index.php/issues/science /item/164-smelling-a-corporate-rat.

3. Stacy Malkan, "Science Media Centre Promotes Corporate Views of Science, " July 20, 2017, usrtk.org/our-investigations /science-media-centre/.

4. Francis Chateaureynaud, "L'étude Seralini, au delà de la polémique," *Transrural initiatives*, no 422, December 2012.

5. *Marianne*, September 29, 2012.

6. Aurélie Haroche, "Les OGM sont probablement moins dangereux pour la santé que les mute news" ("GMOs Are Probably Less Noxious to Our Health Than Mute News") a bit of a strange title for the *Journal International de la Medicine* July 14, 2018.

7. For further details, see Gilles-Éric Seralini and Jérôme Douzelet, *"Poisons cachés ou plaisirs cuisinés"/"Hidden Poisons or Cooking Pleasures,"* Actes Sud, 2014. (*The Great Health Scam*, Natraj Publishers).

8. Called OMICS.

Chapter 5

1. "The Study": This is what they called the research published in *FoodChemTox* by Gilles-Éric Seralini et al., in 2012, republished in 2014.

2. Gilles-Éric Seralini, *Tous cobayes (We Are All Guineapigs)*, Flammarion, 2012.

3. Jean-Paul Jaud, *Tous cobayes? (We Are All Guinea pigs?)* movie based on the book by Gilles-Éric Seralini, DVD with J+B Séquences, 2012.
4. https://usrtk.org/wp-content/uploads/bsk-pdf-manager/2019/02 /Monsanto-motion-to-exclude-Seralini-study-as-evidence.pdf.
5. Eva Novotny, *Journal of Biological Physics and Chemistry* 18 (2018): 32–56.
6. Letter from Monsanto's lawyers to the Supreme Court of California, May 24, 2018. Case n° CGC-16550128.
7. Carey Gillam, "Monsanto Roundup & Dicamba Trial Tracker" July 26, 2021, usrtk.org/pesticides/monsanto-roundup -trial-tracker.

Chapter 6

1. This is the title of a book by Stéphane Foucart, published in 2013 by Denoël.
2. "Envoyé spécial. Monsanto, la fabrique du doute," January 17 2019, www.youtube.com/watch?v=FYnWPcgEqbQ.
3. Monsanto Papers, 2015. MONGLY02063611, MONGLY02063572.
4. International Agency for Research on Cancer.
5. *Food and Chemical Toxicology*, Wallace Hayes, its director, was in an undisclosed contract with Monsanto.
6. *Food and Chemical Toxicology*, Wallace Hayes, its director, was in an undisclosed contract with Monsanto.
7. Johnson and Hardeman verdicts: https://www. baumhedlundlaw.com/documents/pdf/monsanto-documents /johnson-trial/Johnson-vs-Monsanto-Verdict-Form.pdf; https: //www.baumhedlundlaw.com/documents/pdf/monsanto -documents/Hardeman/Order-Denying-Monsanto-Motion -New-Trial.pdf.
8. Reprise Records, 2015. The lyrics of the song are available on https://www.lacoccinelle.net/1003713.html.

9. This concerns studies on animal and human cells, but also on animals and patients.
10. Food and Agriculture Organization of the United Nations, and the World Health Organization.
11. Fabrice Nicolino, *Un empoisonnement universel* (*A Global Poisoning*), LLL (2014): 300–301.
12. MONGLY07018356 (slides revised-redactions).
13. Sophie Richard et al., *Environmental Health Perspectives* 113:6 (2005): 716–720.
14. Research performed with the collaboration of Gérald Jungers: "Toxic compounds in herbicides without glyphosate," http://www.seralini.fr/wp-content/uploads/2020/10/seralini-jungers-2020-food-chem-tox.pdf.
15. Curt T. Della Valle et al., *Cancer Causes Control*, 2016, 27, p. 113; Van Balen et al., *Occupational and Environmental Medicine* 63 (2006): 663–668.

Chapter 7

1. Eva Novotny's title: "Retraction by Corruption: the 2012 Seralini Paper," *Journal of Biological Physics and Chemistry*18 (2018): 32–56. Thesis showing Monsanto's deception in order to retract the article I wrote with my team members.
2. https://www.baumhedlundlaw.com/assets/Monsanto%20Roundup%20pages/Secret%20Documents/monsanto-documents-chart-101217.pdf
3. A French senator was still repeating these stupidities in May 2019, although these had been contradicted for years.
4. See Patrick Moore, former director of Greenpeace who later became a spokesperson for Monsanto, physician, treating the reporter who suggested he drink some Roundup like an idiot; he said it was harmless. https://www.youtube.com/watch?v=Vec_Pgt_86E.

5. Benjamin Sourice, *Mediapart* blog, September 28, 2012, https://blogs.mediapart.fr/benjamin-sourice/blog.
6. *Le Canard enchaîné*, September 26, 2012.
7. Called "F. G." by Gaël Lombart in *Aujourd'hui en France*, January 18, 2019, as he has since passed.
8. Document 24172, 2018, produced in court in California in 2019 during the Hardeman trial, case3:16-md-02741-VC.
9. Ibid.
10. Retraction notice to "Long term toxicity of a Roundup herbicide and a Roundup-tolerant genetically modified maize'" *Food Chem Tox* 63 (2014): 244.

Chapter 8
1. The hearing of January 11, 2017, see "Monsanto Scientist Donna Farmer Deposition (PART 1 OF 2) | Johnson v. Monsanto Co." www.youtube.com/watch?v=k4vAW5fOayQ.
2. This piece of work had already been accepted and the proofs corrected. It was to appear in 2017 in *Plos One*. The publisher then asked me to draft a first version of the press release to help broadcast the results. He needed to share it with the entire editorial staff and the legal department. Soon thereafter, the news arrived: the article would not be published, even though the publication date was a week away. No scientific explanation; an internal decision. It was simply a retraction before publication. It took me almost a year to have it published elsewhere, because the scientific process is a long one. The delay benefited Monsanto.
3. *Hidden Poisons or Cooking Pleasures*, Actes Sud, 2014. Translated in *The Great Health Scam*, by Seralini and Douzelet, Natraj Publisher.
4. Mariano Castro, Carlos Ojeda et Alicia F. Cirelli, "Advances in Surfactants for Agrochemicals," *Environmental Chemistry Letters* 12 (2014): 85–95.

5. CJUE (Court of Justice of the European Union), October 1, 2019, aff. C 616/17, point 75.

6. Hearing of January 11, 2017, see "Monsanto Scientist Donna Farmer Deposition (PART 1 OF 2) | Johnson v. Monsanto Co." www.youtube.com/watch?v=k4vAW5fOayQ.

7. Mark A. Martens et al., *Regulatory Toxicology and Pharmacology* 107:104347 (2019)

8. As the lawyer Kathryn Forgie reminded the European Parliament on September 25, 2017. Also see MONGLY02721133, in which as early as 2010 Monsanto developed a slide show revealing its strategy to hide the toxicity of polyoxyethyeneamine (POEA), (and of surfactants in general) at the international level. Also see MONGLY06414231 and MONGLY03401122.

9. Julie Marc et al., *Chemical Research in Toxicology*, 2002, 15, page 326–331.

10. In France, in the illegal file-collecting scandal wrought by Monsanto's PR firms, Publicis, and Fleishman Hillard, a partner in the worldwide "Let Nothing Go" campaign.

11. France 3 Brittany, May 14, 2019. I comment on the fact that the EFSA was only asked to evaluate glyphosate alone, because the regulation was made to suit the industry.

12. Sixty-three percent of the fosfomycine is absorbed orally, and 37 percent is excreted in the urine of healthy volunteers. Eric Wenzler et al., *Antimicrobial Agents and Chemotherapy*, 2017, 61 (9).

Chapter 9

1. Carey Gillam, "I'm a Journalist. Monsanto Built a Step-by-Step Strategy to Destroy My Reputation," *The Guardian*, August 9, 2019.

2. Stéphane Foucart and Stéphane Horel, "Près de 1 500 personnes ont été fichées par Monsanto en Europe,"

("Monsanto's European operations kept illegal files on approximately 1,500 people"), *Le Monde*, September 7, 2019.

3. See page 23 (final paragraph).
4. https://www.iarc.who.int/featured-news/media-centre-iarc-news -glyphosate/.
5. "Monsanto Spent $17 Million in One Year to Discredit International Cancer Agency over Glyphosate Classification," January, 4, 2019, sustainablepulse.com/2019/04/01.
6. https://www.iarc.who.int/featured-news/media-centre-iarc-news -glyphosate/.
7. https://www.lemonde.fr/planete/article/2019/02/03/nicolas -hulot-se-dit-victime-des-agissements-d-une-officine-belge -chargee-par-monsanto-de-ternir-sa-reputation_5418558_3244. html
8. MONGLY02722492.

Chapter 10

1. Told and referenced in my books *Ces OGM qui changent le monde* (*The GMOs that are changing the world*) and *Génétiquement incorrect* (*Genetically Incorrect*), Flammarion, coll. "Champs," 2004 and 2005.
2. German agency for evaluating pesticides, that played a key role in the EFSA's ruling on glyphosate
3. Pablo Steinberg et al., *Archives of Toxicology* 93 (2019): 1095–1139.
4. Gilles-Éric Seralini, *Environmental Sciences Europe* 32 (2020): 18–24.
5. As I demonstrated in my article that appeared in *Environmental Sciences Europe* in 2020 (vol. 32, no 18), Steinberg's controls were contaminated by glyphosate, other pesticides, GMOs and heavy metals in the food, making the results unusable. In the conclusion I describe in addition his numerous conflicts of interest.

Chapter 11

1. Daniel A. Goldstein et al., *Journal of Applied Microbiology* 99 (2005): 7–23.
2. The editor's notice of retraction: *Food and Chemical Toxicology* 63 (2014): 244.
3. Gilles-Éric Seralini et al., *Food and Chemical Toxicology* 53 (2013): 461–468.
4. The article that *Forbes* pulled following the discovery of Miller's fraud. He had recopied Monsanto's draft. This last was paying him and he didn't declare his conflict of interest: "*Scientists Smell a Rat in Fraudulent Rat Study.*" In the end it will be titled, "*Scientists Smell a Rat in Fraudulent Genetic Engineering Study.*"
5. Document 298810, 03132019, Exhibit 10 at the US court case 3 :16-md-02741-VC.
6. Ibid.
7. The collective industries that produce glyphosate.
8. John Swarthout, September 12, 2019; https://usrtk.org/tag /john-swarthout/ (from Monsanto's lawyers).
9. Document 298810, 03132019, Exhibit 10, case 3:16-md-02741-VC. Supreme Court of California.
10. ensser.org/publications/ensser-comments-on-seralini-et-al-2012/.

Chapter 12

1. Ninety-minute program on Canal+. To watch an excerpt: "Le point de vue de Marc Fellous sur les OGM" www.dailymotion .com/video/xbqnd0.
2. *Conseil supérieur de l'audiovisuel* (*Superior Audiovisual Council*).
3. April 20, 2019, https://www.leparisien.fr/societe/monsanto-et -ses-trolls-aux-racines-du-soupcon-20-04-2019-8057148.php.
4. September 30, 2012.
5. In the trial that he lost.

6. As certified by the police investigation and the judge. He was finally absolved and given the benefit of the doubt. The main witness whose signature was plagiarized was not present at the trial.

7. Foundation of the University of Florida documents, posted by the *New York Times.*

8. USRTK, August 1, 2018, Stacy Malkan's investigation.

9. E. Sautot, Lyon Capitale, 2012, 717, pages 32–36.

10. Paul Deheuvels, *Rebelle-santé*, March 2013, 153, interview pages 34–36, www.seralini.fr/wp-content/ uploads/2019/02/interview -academicien-deheuvels.pdf.

11. "GMA $18 Million Fine Reinstated over Cover Up in GMO Labeling Fight" sustainablepulse.com/2020/04/17/.

12. *La Libre Éco* with *Agence France Presse* April 6, 2020. In June 2020, Bayer settled with 125,000 plaintiffs for $10 billion; 35 000 plaintiffs have still not settled.

13. Dave Goulson, "Pesticides, Corporate Irresponsibility, and the Fate of Our Planet," *One Earth*, v.2.4, p.302–305, April 24, 2020, https://doi.org/10.1016/j.oneear.2020.03.004; see also Carey Gillam, "Court Overturns EPA Approval of Popular Herbicide Made by Monsanto," *The Guardian*, May 1, 2020.

14. Gary Ruskin, "Reporting from Court," February 25, 2019, usrtk.org/monsanto-trial-tacker/reporting-from-court.

15. See Fabrice Nicolino, *Un empoisonnement universel (A Universal Poisoning)*, LLL, 2014, very well-documented regarding Bayer's wartime nazi activities, among other things. The post-war survival of IG Farben in Abwicklung, the chemical manufacturer who produced Zyklon B gas (fine-tuned by its subdivision Degesch, which was used in the death camps) constituted a scandal, including in the eyes of many Germans. Part of IG Farben's chemical activity ended up with Bayer and continued to develop deadly gases, some of which were

derived from Zyklon B: tabun, sarin, soman, and post-war, cyclosarin.
16. As shown in the Weimar Museum near Buchenwald, visited in the fall of 2019.
17. He was still there as of October 10, 2019.
18. Charles-Léopold's book *Plaidoyer pour un contre-lobbying citoyen* (*Plea for a civilian counter-lobbying*), 2014.

Chapter 13
1. Stock, 2006.
2. Hachette littératures, 2009.
3. Included in the *Marianne* letter, September 25, 2012 and for the following on October 5, 2012.
4. This no doubt refers to MON NK 603 which is Roundup-tolerant. He made it possible for this GMO to be authorized (and which I tested). There are many typos.
5. *Actu environnement*, February 25, 2020.
6. "Retraction of Seralini GMO study is attack on scientific integrity," www.endsciencecensorship.org, seen on January 23, 2020.
7. Especially regarding the strain of rat, which was the same one Monsanto used to have their GMO authorized.
8. Gaël Lombart, "Monsanto et ses 'trolls': aux racines du soupçon," *Le Parisien*, April 20, 2019.
9. USRTK, document *AgBioChatter*, September 19, 2012.
10. lalettreducotentin.hautetfort.com. (We won in Court to remove the page.)
11. Investigation by Benjamin Sourice dated May 18, 2015, on over-blog.com.
12. According to Stéphane Horel and Stéphane Foucart, "'Agribashing': un levier d'influence pour une partie du monde agricole," *Le Monde*, February 14, 2020, p. 7.

Chapter 14

1. The Nobel-prize-winner Luc Montagnier's advice "based" on his work in virology was ignored. It explained the manipulations that may have been at the origin of the retrovirus. This should have led to more rapid and efficient international investigations. Just as we could have followed the recommendations of certain scientists for simple treatments. By thus neglecting the advice of health professionals opposed to sweeping stay-at-home orders, these last had nefarious effects on lowered immunity, on the social fabric, and dramatic effects on the population, in particular the most vulnerable. The countries who did not confine people at home, such as Sweden and Hong Kong, had many fewer deaths in the beginning.

2. These information technologies also served to collect personal health data and other data as well, including for the US information services (see Lional Astruc, *L'Art de la fausse générosité* (*The Art of False Generosity*), Actes Sud, 2019, and Edward Snowden, *Mémoires vives*, (*Permanent Record*, Metropolitan Books, 2017; Seuil, 2019). This happened thanks to databases or cell phone chips. Some people went as far as fearing the presence of microchips in vaccines, but were too afraid to speak up, as newspapers belonging to large groups were poised to treat them as conspiracy theorists who were out of touch with the dominant thinking.

3. In 2020 the British company British American Tobacco, that produces among others Lucky Strike, promised a tobacco-based GMO as the basis for a new vaccine against the novel coronavirus. Will tobacco companies be the next saviors of the world? It's true that tobacco cannot be contained. See, Mark Sweeney, "British American Tobacco Working on Plant-Based Coronavirus Vaccine," *Guardian*, April 1, 2020. If they succeed, the vaccine will be extracted from tobacco.

Glossary:
Little Lexicon of Intentional Confusion

*Because it's unbearable for a scientist to see science distorted
into an industrial religion any time our health is at stake.*

Here is a definition of the terms the multinationals use in the most inappropriate ways. The same goes for health and environmental agencies, governments, leaders and authorities, the media, and consequently the NGOs. The result often leads to errors in the language used and misleading of the general public.

Acceptable Daily Intake (ADI): This is calculated for a substance for which they first measure the highest toxic dose that has no observable side effects and is very subjective depending on the methods used to study it. For example, hormonal or nervous disturbances aren't taken into consideration in animals during the evaluation. This supposed nontoxic dose in the short term is usually divided by a safety factor of one hundred. It is classically known that for their homologation the evaluation of commercial products happens in two stages. Initially, only the declared active ingredient is evaluated for the risk to humans and the environment, according to the rules established by the European regulatory agencies (CE number 1107/2009). Thereafter, the product in question (product containing this substance) is evaluated by the national authorities. At this stage there wouldn't be any further long-term tests nor any blood tests, but only reports of the irritation to the skin and eyes that the commercial product provokes. The ADI then, and this is of utmost importance, is calculated based on what the manufacturers declare, in this case glyphosate, evaluated in the same way around the world, and by itself, alone with no other compounds. This is incorrect and constitutes a fundamental scientific mistake. Glyphosate under this distorted method presents a high ADI, meaning that you would have to drink a lot to reach an artificial threshold of toxicity, which is the contrary if you take into account its very toxic formulants.

Active Ingredient: A substance in a pesticide that has been declared active by the manufacturer, and which appears clearly on the label, just as for a medical drug. Its efficacy as a pesticide would not be taken into consideration in a regulatory test as it is a marketing choice. Its toxicity on the other hand should be. Yet the mixture with its formulants is much more toxic and it should always be described as a "declared active ingredient."

Active Principle: Synonym for active ingredient, with the same limits for the reasoning: not proven but only declared by industry, tested alone not with its toxic formulants, especially for long-term risks.

Adjuvants (*adjuvere* in latin): A word used to make people believe that the formulants in pesticides (they are always in combination) simply add a little something or are mere additives to the "active" chemical declared. (see active ingredient – or active principle) This puts them in the same category as water or salt when it is in fact proven that, in the case of Roundup, glyphosate's toxicity is negligeable compared to that of the aforementioned adjuvants, often misunderstood as simple helpers. You could apply the same criterion to most pesticides including Roundup and other herbicides without glyphosate, or to neonicotinoids, or SDHIs [1] Succinate

Dehydrogenase Inhibitors (inhibitors of cellular respiration), as well as to vaccines and certain medications. It is the responsibility of the manufacturer requesting commercial authorization to declare the category to which the substance belongs for regulatory purposes (i.e., whether an ingredient is "active" or rather an adjuvant which will only be tested over a short period, not in the long term). The pesticide or GMO's efficacy isn't taken into consideration during the evaluation of its toxicity: In medicine an adjuvant to a medication is most often used to delay its intestinal or renal elimination (capsule in cellulose, lactose . . .) or to avoid its too rapid penetration; a pesticide adjuvant kills cells all by itself by opening pores, by burning and penetrating, just as it does through cuticles (smooth surface, or more rigid) of insects or plants.

Co-formulants: Another name that helps avoid the word "adjuvants," but that puts an emphasis on the ingredients deemed active because that would indicate that it is "co-acting." If the stated active ingredient isn't actually that, like glyphosate acting in the midst of much more toxic mixtures than it (we proved this many times), the term "formulants" instead is always correct.

Diluents: Another name for adjuvants, serving to make believe in their innocuity and to divert attention from the active ingredient listed. For example, these so-called diluents multiply the toxicity of glyphosate by a factor of one thousand, while companies make it seem from the word that they dilute it.

Dispersants: There aren't any hidden poisons with more names in the oversight of the supposed "inerts." This is another one. These agents combined with the active ingredients listed in pesticides supposedly "disperse" them on plants, therefore in the environment. Due to their oily nature, they can, on the contrary, accumulate in living organisms together with the listed active ingredients.

Emolients: A synonym for "diluents" or "humectants" and an inadequate term for talking about the "formulants" as if they were not toxic. What "wets" mostly evokes water.

Emulsifiers: In theory, these are agents that permit oily and aquas substances to stay together. But this word is also used to qualify the compounds in the so-called "inactive" formulants in pesticides. While it is true that ingredients such as lecithin, that can be extracted from plants, are inactive, the petroleum derivatives accompanying them are less eatable!

Formulants: an appropriate name to suggest the mix of products in a pesticide, added to a listed active ingredient. Contrary to the belief of the regulatory agencies, a pesticide is not marketed as a molecule purified in water, but always with its formulants. We identified a much higher toxicity in them due to petroleum derivatives and heavy metals.

GMOs Containing Pesticides or Pesticide-Rich GMOs: These include GMOs that tolerate one or more pesticides, a characteristic of agricultural GMOs, such as soy or corn with Roundup. These GMOs absorb the pesticides, therefore, they can contain a high dose without dying. They have been genetically modified to do this. The second category includes GMOs that produce one or more insecticides within their cells (therefore we eat them), and they leak into the earth.

Glyphosate: The myth generated by Monsanto suggests that their leading herbicide is glyphosate, a rather simple molecule that resembles an amino acid. In actual fact, it is paired with very toxic formulants, that have not been tested on organisms over the long term. This was declared illegal by the Court of Justice of the European Union in the last trimester of 2019. The myth made it possible for Monsanto to have glyphosate evaluated on its own by Europe and other continents, hence grossly underestimating the toxicity of the pesticide.

GMO Containing Pesticides or Pesticide-Rich GMOs: These include GMOs that tolerate one or more pesticides, a characteristic of agricultural GMOs, such as soy or corn with Roundup. These GMOs absorb the pesticides, therefore can contain a high dose without dying. They have been genetically modified to do this. The second category includes GMOs that produce one or more insecticides within their cells (therefore we eat them), and they leak into the earth.

Historical Data: In the safety tests of agrochemical and pharmaceutical products, their use is a way to mask inconvenient side effects of the tested products. These are blood and urine tests on so-called "normal" rats, obtained from tests performed by the industry decades ago (hence their historical nature) and still today. These animals were fed croquettes that today were proven highly contaminated, and therefore, add very faulty values for multiple comparisons. These animals can develop chronic illnesses and thus present disturbed blood and organ parameters that could be taken erroneously as normal. In these conditions, the rats usually have many tumors and are very sick: companies conclude that 71 percent of them have normal or spontaneous breast tumors, and 90 percent have pituitary tumors, to take a few organs as examples! They then deduce that this is all linked to the species or to the strain of rat used. I've been criticized for not taking this historical data into account in my 2012 study. I did this on purpose in order to obtain more realistic results. And I had specially made pesticide-free croquettes for the control group.

Humectants: A synonym of "wetting agents" or "emolients" or "diluents." An inappropriate term for the same reasons.

Inerts: By employing this kind of word for formulants, the industry and those who support it make the public believe that the additives are harmless, when the contrary is true.

Pesticides: Insecticides and herbicides are the two main categories of pesticides, along with fungicides, raticides, etc. They are marketed mixed in with their formulants.

Phytosanitary: An inappropriate publicity term, created by the industries to avoid the scientific term "pesticide" and the bad press it generates due to the many side effects observed. In reality, the etymological meaning of "phytosanitary" is "plant curing," when in fact herbicides for the first category of pesticides kill them. In the same way insecticides are supposed to kill insects who eat the crops before we do, and not for nurturing plants. The industry is trapped in its own public relations lexicon and so most often tries to use the term "phytopharmaceutical" in its place.

Resistance: An inadequate term for agricultural GMOs. The industry evokes "resistant to herbicides" as if the modified plant rejects them or fights against them. In fact, it accommodates, tolerates, and absorbs them. "Resistant" GMOs are not strong plants that would battle pesticides. In GMO soy with Roundup for example, the molecular tolerance phenomenon is passive, and doesn't act like a resistance. To better understand, tolerance and resistance to the Nazis are two contrary acts that are not equal. One should say of a Roundup tolerant GMO and not resistant to Roundup. For the other GMOs, "resistance to insects" is evoked, when the plants in question are modified to release one or more of the insecticides that are supposed to poison certain types of insects. The term "GMO insecticides" is more appropriate in this case.

Surfactants: Inadequate term used to make it appear that the formulants in pesticides reside on a plant's surface (which would limit its toxicity when we eat it). In reality, they are very penetrating, and are composed of corrosive oxidized products, petroleum derivatives and heavy metals.

Tensioactive Properties (or Agents): Used as an improper synonymous of "surfactants," these compounds are supposed to be agents remaining on cell surfaces. In fact, these are penetrating agents like other formulants.

List of Principal Characters in the Book

The Monsanto Staff

Donna Farmer, chief toxicologist for Monsanto, later for Bayer.

Harvey Glick, in charge of regulatory and scientific affairs for Asia.

Daniel Goldstein, pediatrician and clinical toxicologist. Product regulation advisor.

Bruce Hammond, toxicologist who works to hide the toxicity of the GMOs.

Charles Healy, toxicologist.

William Heydens, specialist in psychiatry and serves as a toxicology biologist.

Dan Jenkins, in charge of relations with the US government.

Eric Sachs, one of the figureheads who seems to supervise the lot of them.

David Saltmiras, "new chemistry toxicology lead," Wallace Hayes's former student (see the next section).

David Stark, vice president.

John Swarthout, in charge of controversial topics, moved over to Bayer.

John Vicini, responsible for the rewards for going against Seralini, moved over to Bayer where he takes care of scientific affairs for food safety.

Those Who Help the Chemical and GMO Industries

John Acquavella, Professor in the epidemiology department at the University of Aarhus (Denmark), on Monsanto's payroll.

Jean-François Bach, member of the French Academy of Sciences, who defended its position against Seralini.

Francine Casse, former member of the Biomolecular Engineering Commission (CGB) and of the French Association of Plant Biotechnology (Association française des biotechnologies végétales), and one of Axel Kahn and Gérard Pascal's first assistants.

Bruce Chassy, Professor Emeritus of Food Science and Human Nutrition currently retired from the University of Illinois, often on Monsanto's payroll.

Daniel Dattée, beet farmer who labors for the acceptance of GMO beets.

Yvette Dattée (Daniel's wife), member of the CGB and the French Association of Plant Biotechnology (Association française des biotechnologies végétales).

Roland Douce, member of the American Academy of Science and the French Association of Plant Biotechnology who played a crucial role in a report supporting GMOs.

Marc Fellous, sentenced for defamation in 2011, specialist of the Y chromosome and former (chief expert of plant GMOs) legendary for his prose.

Kevin Folta, Professor of the Horticultural Sciences at the University of Florida and then at the University of Illinois. On Monsanto's payroll, then on to Bayer, otherwise passionate about research on strawberries.

François Gendre, Director of life sciences for Danone, member of the ILSI. He served as a relay between Monsanto and the French government at the time the 2012 study appeared.

Richard Goodman, Professor at the University of Nebraska who worked for Monsanto. In 2012 after the study appeared he became the head of biotechnology at the review that published and then was the one to retract it.

Wallace Hayes, The compromised editor in chief of *Food and Chemical Toxicology* who published the study in 2012 and then, with the help of Richard Goodman retracted it in 2013. He was in an undisclosed contract with Monsanto.

Louis-Marie Houdebine, renowned biotechnology expert of fluo-engineered rabbit fame.

Axel Kahn, former president of the CGB and founding member of a pro-GMO association (The French Association of Plant Biotechnologies). Has more recently written books on hiking. Former media favorite on questions of genetics.

Kate Kelland, from Reuters, a relay for Monsanto.

Marcel Kuntz, member of AgBioChatter, scientist who organized the bad-mouthing of Seralini.

Henry Miller, Nom de plume for Monsanto, and former cheerleader of the firm, driver of a red sedan.

Wayne Parrot, Professor of Crop Sciences at the University of Georgia, from Kentucky, member of AgBioChatter.

James Parry, Welsh scientist who was under contract with Monsanto but was released over his inconvenient findings.

Gérard Pascal, nutritionist who touted the safety of GMOs. European pro-biotechnology activist who prefers to remain in the shadows.

Georges Pelletier, member of the French Academy of Sciences, in charge of biotechnology.

Wackes Seppi, pseudonym for André Heitz, courageous blogger.

Pablo Steinberg, member of the International Life Sciences Institute with numerous conflicts of interest. Selected by the EFSA to redo Seralini's 2012 study (his study was not performed on Roundup but instead of a corn GMO alone, using very contaminated controls). Gilles-Éric Seralini published an extensive study in 2020 to refute Steinberg's and to argue for its retraction.

Anton Suwalki, pseudonym for Stéphane Adrover, courageous blogger.

Alain Toppan, GMO manufacturer.

Marc Van Montagu, one of the first manufacturers of GMOs, in particular a transgenic tobacco, not in any way to feed the world but to contain an antibiotic-resistant gene.

Gil Rivière-Wekstein, takes himself for a journalist. Hosts of the site *Agriculture & environnement*. He was sentenced for defamation as early as 2007.

The Sick, the Plaintiffs, and their Lawyers

Michael Baum, partner at Baum Hedlund law. Played a key role in obtaining the Monsanto Papers and in bringing about the court cases against Monsanto.

William Bourdon, Gilles-Éric Seralini's lawyer in the new cases against reporters and Monsanto in 2019, not yet sentenced.

Apolline Cagnat, associate of William Bourdon.

Bernard Dartevelle, lawyer who helped Seralini win the seven defamation cases mentioned in the work: (1) against Marc Fellous, (2) against the *l'Association française des biotechno-logies végétales* (The French Association of Plant Biotechnologies (3) against the magazine *Marianne* in the first instance (4) and on appeal (5), against the reporter Jean-Claude Jaillette in first instance (6) and on appeal (7) against the blogger Daniel Dubost.

Cindy Gay, associate of Bernard Dartevelle.

Dewayne Johnson, the first patient to win a case against Monsanto.

Kathryn Forgie, one of the first lawyers in the investigation, connected to Michael Baum's law firm.

Edward Hardeman, patient, plaintiff in the second case in California that was won against Monsanto.

Leemon McHenry, research consultant with Baum Hedlund who analysed the Monsanto Papers, and among other things, coauthor of *The Illusion of Evidence-Based Medicine* (Wakefield Press, 2020).

The Pilliods, husband and wife, both patients, were the plaintiffs in the third historic lawsuit against Monsanto (in which Monsanto was fined $2 billion in the first instance.

Guillaume Tumerelle, lawyer who fought and won the case for an adequate evaluation of the toxicoloy of the formulants in Roundup.

Andrew Wagstaff, renowned lawyer for plaintiffs against Monsanto and author of reports to force the courts to take our studies into account.

Brent Wisner, renowned lawyer from the Baum law firm, won lawsuits against Monsanto.

Seralini's Supporters

Robert Bellé, discoverer of the toxicity of Roundup on sea urchin eggs, a cellular model.

Charles Benbrook, author of a report showing Monsanto's dishonesties against Gilles-Éric Seralini. Director from 1984 to 1990 of the commission on agriculture at the American Academy of Sciences. He recently published on the formulants in Roundup and their toxicity together with a former graduate student of Gilles-Éric Seralini's, Robin Mesnage.

Theo Colborn, deceased renowned American environmentalist. Gilles-Éric Seralini is a recipient of the prize in his name.

Paul Deheuvels, member of the *Académie des sciences* (French Academy of Sciences) who denounced the underhandedness of this instance, witness in Seralini's lawsuits.

José Luis Domingo, scientist, original editor of the 2012 article in *Food and Chemical Toxicology*.

Francis Hallé, renowned scientist and expert in biodiversity.

Albert Jacquard, renowned scientist.

Yvon Le Maho, member of the Academy of Science and course administrator at the *École nationale d'administration* (National

School of Administration), to which he invited Seralini. Witness at one of Seralini's lawsuits.

Jean-Marie Pelt, scientist and one of the most important francophone nature writers. He and Gilles-Éric Seralini were very close (he died in 2015) and coauthored a book together.

Sophie Richard, scientist, graduate student member of Seralini's team and one of the first to be attacked.

Jacques Testart, scientist, one of the scientists behind the first test-tube baby project.

Louise Vandelac, renowned sociologist of Science at the University of Québec in Montreal.

The Investigative Journalists

Carey Gillam, of the *Guardian*, is in charge of US Right to Know (USRTK), which ensures the continuation of the court cases against Monsanto.

Hervé Kempf, former reporter for *Le Monde*, editor-in-chief of *Reporterre*.

Gaël Lombart, of the *Parisien aujourd'hui*.

Élise Lucet, reporter for among others the show *Envoyé spécial* (France 2).

Fabrice Nicolino, from *Charlie Hebdo*, author of numerous works and stories on food and the environment.

Gary Ruskin, cofounder and codirecteur of US-RTK.

Benjamin Sourice, freelance.

Tristan Waleckx, documentary filmmaker for *Envoyé spécial* (France 2).

The Politicians

Claude Allègre, former minister of research in France who subsidized biotechnology companies, climate-change denier, slandered Gilles-Éric Seralini.

Nicolas Hulot, Minister of State in France for Ecological Transition and Solidarity (he resigned), famous environmentalist who sadly was never able to lend his voice as to the toxicity of the formulants of Roundup.

Ségolène Royal, former Minister of the Environment in France who banned formulations of Roundup (containing chemical substances of the POEA family).

Documents of The Seralini Affair

The following documents are accessible via the QR code below.

Discovery
The heartfelt cry of the popular jury against Monsanto during the Johnson trial.

The Explosion of 2012
Monsanto's nom de plume Henry Miller, in *Forbes*.

It's True Because We Say So
Motion by Monsanto's lawyers to eliminate all discussion of Seralini's work from the jury's deliberations. This type of motion was introduced in every trial on the subject in the United States.

The lawyers for the plaintiffs used this argument in order to incorporate the indisputable evidence revealed in Seralini's work.

The Making of a Lie
Monsanto is punished for dishonesty by the US courts and sentenced to pay the historic sum of $2 billion. Meanwhile, the rest of the world remains deaf to the dangers of Roundup.

Retraction by Corruption
The Judas of science, who discredits it to the tune of $400 an hour.

The Dark Side of the Shadow Army
Donna Farmer's testimony that Roundup, on the market since 1974, was not tested.

"A Pain in the Ass"
Elegance according to Monsanto: Seralini is a "persistent bugger."

Fate or Bulgarian Umbrella?
The prizes Monsanto offered for actions against Seralini while he is in London Hospital.

Seralini against the "Scientific Community"
In which the president of a so-called independent governmental commission on GMOs (the CGB), reports back to Monsanto that he lost his trial against Seralini, but that he knows that the "precious support" of the company will endure.

Toxic Information: The Infiltration of the Media
In which the very same president who authorized the GMO asked Monsanto (in very bad English) whether they had tested the product which was the focus of Seralini's study. He didn't know.

A relevant but not valiant Turkish scientist: if the type of rat Seralini used was not the right kind, then the authorization for GMOs must be withdrawn.

By opening this link, you will access a selection of the Monsanto Papers as well as the actual court documents that led the US justice system to condemn Monsanto during the trials mentioned in the book, as well as some of the more relevant documents showing the company's dishonesty.

Acknowledgments

Thank you for your attention to the numerous names mentioned in the book. Despite their zeal at the service of the system, they are for the most part only cogs in the wheel, with the exception of the leadership and their billionaire cheerleaders. And even then, there are many more names of people who fight actively against our studies; the list here is not exhaustive.

In this volume we were of course not able to analyze the 55,952 documents that discuss Gilles-Éric Seralini in the Monsanto Papers. May those who worked to research them, to read, classify, and advise us on which ones to choose, be wholeheartedly thanked. We're referring to Michael Baum, Brent Wisner, and especially, the wise Leemon McHenry. Carey Gillam's analyses as well as those of Gary Ruskin from the US Right to Know were essential. Bernard Dartevelle played a crucial role and was a remarkable investigative attorney, as was his associate Cindy Gay, during the seven trials described. The attorneys William Bourdon and Apolline Cagnat support and encourage us on for the trials yet to come. Maître Guillaume Tumerelle and Aude Desaint played a major role in getting our work on the importance of having the formulants of Roundup evaluated and recognized at the European Court of Justice, and we are grateful to them.

Books by the Same Authors

Gilles-Éric Seralini [Extracts]
L'évolution de la matière (The Evolution of Matter)

Books
Ces OGM qui changent le monde (These GMOs That Change the World)
Nous pouvons nous dépolluer (We Are Able to Detox Ourselves)
Tous Cobayes (We Are All Guinea Pigs)—[adapted for a movie]

Coauthored with Jérôme Douzelet
L'Affaire Roundup à la lumière des Monsanto Papers (The Monsanto Papers)—[French original]
Le goût des pesticides dans le vin (The Taste of Pesticides in Wine)
Plaisirs cuisinés ou poisons caches (Culinary Pleasures or Hidden Poisons)
The Great Health Scam